LAND USE CHANGE IN TROPICAL WATERSHEDS

Evidence, Causes and Remedies

LAND USE CHANGE IN TROPICAL WATERSHEDS
Evidence, Causes and Remedies

Edited by

Ian Coxhead

Department of Agricultural and Applied Economics
University of Wisconsin-Madison
USA

and

Gerald Shively

Department of Agricultural Economics
Purdue University
Indiana
USA

CABI Publishing

CABI Publishing is a division of CAB International

CABI Publishing
CAB International
Wallingford
Oxfordshire OX10 8DE
UK

Tel: +44 (0)1491 832111
Fax: +44 (0)1491 833508
E-mail: cabi@cabi.org
Website: www.cabi-publishing.org

CABI Publishing
875 Massachusetts Avenue
7th Floor
Cambridge, MA 02139
USA

Tel: +1 617 395 4056
Fax: +1 617 354 6875
E-mail: cabi-nao@cabi.org

A catalogue record for this book is available from the British Library, London, UK.
A catalogue record for this book is available from the Library of Congress, Washington, DC.

Land use change in tropical watersheds: evidence, causes, and remedies / edited by Ian Coxhead, Gerald Shively.
    p. cm.
Includes bibliographical references.
ISBN-13: 978-0-85199-912-8 (alk. paper)
ISBN-10: 0-85199-912-3 (alk. paper)
1. Land use, Rural--Philippines--Case studies. 2. Watershed management--Philippines--Case studies. I. Coxhead, Ian A. II. Shively, Gerald. III. Title.

HD908.L36 2005
333.76'09599--dc22

2005011812

ISBN 0 85199 912 3
978 0 85199 912 8

Typeset by SPI Publisher Services, Pondicherry, India.
Printed and bound in the UK by Cromwell Press, Trowbridge.

Contents

The colour plate section can be found following p. 132

Contributors

Ian Coxhead, Department of Agricultural and Applied Economics, University of Wisconsin-Madison, 427 Lorch Street, Madison, WI 53706, USA; e-mail: coxhead@wisc.edu

Antonio Daño, Ecosystems Research and Development Bureau (ERDB), College, Laguna 4031, The Philippines; e-mail: tonydanolb@yahoo.com or erdb@laguna.net

Marian delos Angeles, Environment Department, The World Bank, 1818 H Street NW, Washington, DC 20433, USA; e-mail: mdelosangeles @worldbank.org

Bayou Demeke, Department of Agricultural and Applied Economics, University of Wisconsin-Madison, 427 Lorch Street, Madison, WI 53706, USA; e-mail: demeke@aae.wisc.edu

William Deutsch, International Center for Aquaculture and Aquatic Environments, Department of Fisheries and Allied Aquacultures, Auburn University, Auburn, AL 36849, USA; e-mail: deutswg@auburn.edu

Victor Ella, Land and Water Resources Division, Institute of Agricultural Engineering, College of Engineering and Agro-Industrial Technology, University of the Philippines at Los Baños, College, Laguna 4031, The Philippines; e-mail: vbella@up.edu.ph

Kwansoo Kim, Department of Agricultural Economics and Rural Development, Seoul National University, Gwanak Campus Bldg 200, Seoul, Republic of Korea; e-mail: kimk@snu.ac.kr

David Midmore, School of Biological and Environmental Sciences, Central Queensland University, Rockhampton, QLD 4702, Australia; e-mail: d.midmore@cqu.edu.au

Todd Nissen, Bureau of Economic and Business Affairs, U.S. Department of State, 2201 C Street, NW, Washington, DC 20520, USA; e-mail: nissentm@state.gov

Jim Orprecio, 550 Mehan Street, Parkshomes Subdivision, Muntinlupa City, The Philippines; e-mail: jorprecio@amore.org.ph

Stefano Pagiola, Environment Department, The World Bank, 1818 H Street NW, Washington, DC 20433, USA; e-mail: spagiola@worldbank.org

Eduardo Paningbatan Jr, Department of Soil Science, University of the Philippines at Los Baños, College, Laguna 4031, The Philippines; e-mail: dss@laguna.net

Durga Poudel, Department of Renewable Resources, University of Louisiana, Lafayette, LA 70504, USA; e-mail: ddpoudel@louisiana.edu

Agnes Rola, Institute of Strategic Planning and Policy Studies, University of the Philippines at Los Baños, College, Laguna 4031, The Philippines; e-mail: arola@laguna.net

Gerald Shively, Department of Agricultural Economics, Purdue University, 403 West State Street, West Lafayette, IN 47907, USA; e-mail: shivelyg@purdue.edu

Charles Zelek, US Department of Agriculture, Natural Resources Conservation Service, 14th and Independence Ave., SW, Washington, DC 20250, USA; e-mail: chuck.zelek@usda.gov

Grant Zhu, School of Biological and Environmental Sciences, Central Queensland University, Rockhampton, QLD 4702, Australia; e-mail: g.zhu@cqu.edu.au

Preface

High rates of hillside erosion and downstream sedimentation are among the most important agricultural externalities in the developing world. Suspended solids reduce the quality of water for human consumption. Silting of streams increases the risk of flash floods. Accumulation of silt in coastal habitats reduces productivity of aquatic ecosystems and accumulation of sediment in reservoirs reduces hydroelectric capacity and accelerates wear on turbines and power-generating equipment. Of additional concern is that erosion from upland farming creates sediment in downstream irrigation systems, reducing both their productivity and expected life. The last of these concerns is especially important in light of evidence that a lack of reliable water supply precludes expansion and intensification of agriculture in many low-income areas of Asia. In response to these and other environmental concerns, the protection and management of watersheds has become a policy imperative throughout the developing world, and especially in densely populated regions of South and South-east Asia.

The importance of studying watersheds in an integrated way, using an approach that combines biophysical and socio-economic perspectives, is well established. Researchers engaging in watershed-level analysis have emphasized physical characteristics of upstream externalities, and in some cases have made important progress in modelling the main hydrological processes associated with downstream damages. In this book our colleagues and we take up the task of understanding processes leading to watershed degradation by examining the local and national policy environment in which decisions regarding land use are made. Our study focuses on the Manupali River watershed in the Philippine province of Bukidnon. The research described here encapsulates studies that have been conducted over more than a decade and includes examinations of biophysical, economic and social processes and their interconnections. Our analysis explicitly accounts for the fact that, in the Manupali River watershed, as in many tropical watersheds, rates of land degradation are strongly influenced by the decisions of upland farmers. These decisions, in turn, are influenced by the interplay of wages, prices, institutions and opportunities in the general economy.

The work presented in this book provides important generalizable insights into the drivers of environmental outcomes in rural and frontier areas of developing economies. Our efforts build on a large and growing body of empirical research undertaken at the microeconomic, sectoral and general equilibrium levels. Much of this economic foundation comes from research (including extensive fieldwork) conducted in the watershed. Our findings show that the policy context of land degradation is complex and involves economic incentives and constraints that are determined by local, provincial and national level policymakers. An important objective for policymakers is to find ways to promote sustainable agricultural activities in sloping and upland areas, many of which provide unvalued or severely undervalued ecosystem services. These environments are typically characterized by ecological richness but human poverty. As prior studies have shown, land degradation resulting from high rates of erosion presents an especially significant problem to watershed communities. The main story that emerges from this book is that rates of upland erosion can often be traced to specific economic policies that either encourage the planting of erosive crops or discourage the adoption of more environment-friendly crops. For the policy analyst, therefore, our findings suggest that policy research should account for the impacts of economic policies over time and the interactions between economic actors and their physical and policy environment.

The research in this book contributes to a growing body of work that seeks to merge studies of economic processes with those examining environmental processes. Our level of analysis is the watershed, and the Manupali, like many tropical watersheds, is characterized by an aggregation of various agricultural and non-agricultural land uses. Our aim is to assist in the evaluation of social, economic and environmental implications of changes in public policy, especially changes that affect the ways in which land and other natural resources are used. The book has two specific objectives: to provide a framework for studying economic and environmental impacts of policy changes, especially as they impact upland agricultural producers and downstream communities; and to document through the example of the Manupali River watershed the empirical forces leading to land use change and the potential impact of policy changes. We hope that our approach and findings will serve as guides for natural resource and environmental research and planning, both by academics and by policymakers. Our work describes and seeks to understand the key interrelationships between biological, economic and social phenomena, so as to support improved decision making at local, national and international levels.

All the research presented in this book was conducted as part of the South-east Asia programme of the Sustainable Agriculture and Natural Resources Management Collaborative Research Support Program (SANREM CRSP). The research was made possible through support provided by the Office of Agriculture and Food Security, Bureau for Global Programs, US Agency for International Development, under the terms of Award No. PCE-A-

00-98-00019-00. We especially thank Bob Hedlund of USAID, former Chief Technical Officer for the SANREM CRSP, and Carlos Perez of the University of Georgia, former SANREM CRSP project manager, for their support of our efforts.[1] Many other individuals contributed in important ways to the creation of this book. Dr Agnes C. Rola, our research collaborator and co-author, and her staff at the University of the Philippines–Los Baños Institute for Strategic Planning and Policy Studies, have been critically important participants at every stage, from research design to data collection and analysis to brainstorming over interpretation of results. Dr Victoria (Vicky) Espaldon and Ms Doracie Nantes of the Department of Geography, University of the Philippines–Diliman, hosted a research workshop in January 2004 at which many of the research papers contained in this book were presented. Kathryn Boys and Priya Bhagowalia carefully and constructively edited many of the chapters at Purdue. Tim Hardwick, our editor at CABI, was especially supportive and patient throughout the completion of the book. Finally, our families were especially understanding and tolerant as this manuscript consumed many evening and weekend hours that rightfully belonged to them, a sacrifice for which we are very grateful.

Ian Coxhead   Gerald Shively
Madison, Wisconsin   West Lafayette, Indiana

---

[1]The reader's attention is drawn to two other volumes from the same project: Rhoades, Robert (ed.) *Development with Identity: Community, Culture and Sustainability in the Andes,* CAB International, Wallingford, UK; and Moore, Keith (ed.) *Conflict, Social Capital and Managing Natural Resources: a West African Case Study,* CAB International, Wallingford, UK.

1 Economic Development and Watershed Degradation

I. COXHEAD[1] AND G. SHIVELY[2]

[1]Department of Agricultural and Applied Economics, University of Wisconsin-Madison, 427 Lorch Street, Madison, WI 53706, USA; e-mail: coxhead@wisc.edu; [2]Department of Agricultural Economics, Purdue University, 403 West State Street, West Lafayette, IN 47907, USA; e-mail: shivelyg@purdue.edu

Agricultural Development and Environmental Change

In recent decades the protection and management of watersheds has emerged as both a local and national policy imperative throughout the developing world, especially in densely populated regions of South and South-east Asia (World Bank, 1992; Asian Productivity Organization, 1995; Heathcote, 1998).[1] Throughout Asia, growing populations and mismanagement of complex, fragile and poorly understood ecosystems continue to jeopardize the livelihoods of local populations as well as the prospects for environmental protection and rehabilitation. Moreover, across many tropical landscapes, areas of environmental concern closely coincide with pockets of poverty. The biophysical links between upland farms and downstream water users have been well documented for some time. However, policies to promote sustainable use of watersheds populated mainly by the very poor remain elusive.

In this book we ask how economic development, and in particular agricultural development, influences watershed health. We define economic development to mean both economic growth and institutional evolution. Our empirical focus is on a single agricultural landscape, the Manupali River watershed in southern Philippines; however, lessons from this location apply well beyond the borders of the watershed, or indeed the Philippines or South-east Asia. The contributions to this book aim to improve understanding of the general forces influencing patterns of land use in tropical agricultural landscapes. The interface between human activity and the local environment is of central concern, but we also attempt to answer questions of how best to manage natural resources so as to sustain agricultural livelihoods, and how best to minimize environmental damage arising from agricultural activities.

For more than a decade, the standard references on watershed-related themes in Asia have been the volumes edited by Easter et al. (1986) and Doolette and Magrath (1990). These books remain important, not least for having drawn attention to watersheds as 'units of analysis' in their own right. Doolette and Magrath (1990: 2) pointed out that:

> watershed problems are . . . connected by the fact that they can best be understood and dealt with in the context of physical planning units defined by the flow of water.

Similarly, Dixon and Easter (1986: 9), in their introductory chapter to the book edited by Easter *et al.* argued that:

> watersheds, analyzed and managed in an integrated manner, are a useful tool for planning and implementing rural development efforts . . . [t]he question . . . is how the resources within a watershed should be used to obtain a socially optimal level of production over time.

This focus on the watershed as a proper unit of analysis for questions relating to agricultural development has served as an organizing principle for much of the research undertaken since (e.g. Munasinghe, 1992; Naiman, 1995). The landscape orientation was, in fact, a cornerstone for the larger project of which the research reported in this book is a part.

Despite widespread acceptance of the watershed as a unit of analysis, over the past decade several additional themes have emerged that demand a re-examination of the forces shaping land use in tropical watersheds and suggest the need to update the development dialogue of the early 1990s. First among these is the often rapid economic integration of watershed economies with the national economies in which they exist, especially as this integration arises through the maturation of transport and telecommunications infrastructure and the growth of input, product and domestic labour markets. Second, globalization has meant that international trade patterns now often strongly influence patterns of agricultural production in tropical watersheds. Third, the process of decentralization has expanded the scope for local political jurisdictions to shape patterns of economic development – for better or for worse. And fourth, civil society (often through non-governmental organizations) can increasingly be found participating in the decision making process.

A second force shaping our work is the rapid advance in both the number and sophistication of available modelling tools. While technological constraints meant that earlier studies could only point to the need for GIS-based tools and integrated economy–environment models, we can now report results obtained using these. Moreover, because the contributors to this book have been engaged in data collection in one site for more than a decade, we have an unprecedented amount of high-quality data (especially biophysical and panel survey data) available to study watershed-scale phenomena. These data lead us to conclude that many of the problems and issues relevant to watersheds transcend their physical boundaries. In other words, understanding the flow of water is rarely sufficient for understanding watershed health. People, goods and ideas all flow into and out of watersheds, and this necessitates a view of the watershed both as an integrated biophysical system and as part of a much larger economic system.

It is exactly such a view that we hope to promote in this book, which as a whole seeks to illuminate connections between human and natural systems as they exist in both space and time. At a conceptual level, these connections include an interlinked set of biophysical, economic and social features that evolve in response to human desires and needs on the one hand, and to environmental health and resilience on the other. At a more practical level, one recurrent theme in the book is that both the magnitude and the strength of human–environment connections are influenced by a broad range of factors (e.g. market signals and institutional performance) that are shaped by local, regional and national policies. Although it is clear that physical and economic outcomes observed on small farms result from household decisions and the physical features of the landscape, it is the primary thesis of this book that a broader perspective is required. Local and immediate outcomes reflect prevailing conditions in markets in which households participate. Crops and cropping systems are chosen by households from a set constrained, in part, by the choices and trade-offs made by national governments and international markets and organizations. And patterns of innovation, adoption and adaptation reflect the social fabric of which rural households are a part. Decisions made by smallholders can be traced to local economic conditions, and these local conditions, in turn, can be seen to be filtered through local institutions, local rules and social norms. But local economic conditions are also products of national and international economic signals, and are shaped by the performance of regional markets.

The enabling and conditioning factors influencing watershed outcomes include global economic conditions, national government policies and regional market performance. When combined with a standard list of local factors, such as local economic conditions, biophysical features of the landscape and farm-level constraints on activity, this list suggests not only multiple points of entry for researchers and policymakers concerned with patterns of agricultural land use, but also the need for a multi-disciplinary perspective to understand how decisions and outcomes are influenced by economic, social and biophysical features of a landscape. The relationships and linkages of interest in this book are those spatial and temporal connections among economic incentives, social and biophysical conditions and environmental outcomes.

Most of the chapters that follow examine land use changes retrospectively, asking how features of the landscape have evolved in response to key variables of interest. But we also undertake the more difficult task of looking ahead, asking how the agricultural landscape might change in response to changes in policies, institutions or economic signals. Throughout, we maintain our focus on the watershed as a unit of analysis, utilizing a number of techniques to explain and predict land use changes and outcomes at this scale. These approaches include systems-based models, process-oriented simulation models, statistical analysis and GIS-based methods, and participatory techniques that have involved local communities in the measurement and understanding of land use changes in their communities.

Geographical and Historical Context for the Book[2]

The setting for this book is a watershed in the uplands of South-east Asia. These uplands present a rich and diverse natural environment. Over the past 50 years, however, they have been subject to timber extraction, intensified subsistence farming, plantation establishment and other commercial activities that increasingly include highly capital-intensive horticulture and livestock-rearing operations. The expansion and commercialization of agriculture and rural industries has been driven by a combination of demand factors (especially the search for land and forest as complements to labour), supply factors including unregulated access to many resources, and reduced transaction costs due to the expansion of roads and other infrastructure. Without the necessary policy and institutional support for environmental management, these activities have denuded forest cover, polluted rivers, eroded soils and diminished biodiversity. In many areas, deterioration of the natural environment has reached the point at which the viability of future production activities on the same resource base is in question.

Case studies of resource depletion associated either with rapid economic growth or with poverty-driven environmental degradation (e.g. Perrings, 1989) abound in the literature. Many of these identify 'institutional failures' that have created open access to natural resources as an enabling condition. A longer time ρ rspective, however, reveals that open access was not always the norm in uplands; nor, if the experience of comparable areas in wealthy countries is a guide, does this institutional failure persist as economies develop. Clearly, the ways in which economic growth interacts with institutions and influences their evolution are potentially very important determinants of the uses of natural resources and thus (through feedback mechanisms) of the health of upland economies themselves.

Historically, patterns of economic and agricultural development in upland areas have been accompanied by distinct phases of institutional development. We illustrate these phases of agricultural and institutional co-evolution in Fig. 1.1, and provide additional details regarding their characteristics in Table 1.1. In the first phase, prior to colonization or massive internal migration, upland populations are sparse and production is primarily for subsistence, and natural resource use is governed by local customary laws. Tribes and communities manage the resources over which they have control. These institutions are effective as means to govern the commons, not least because demand for the resources is low, technologies for their exploitation are limited, and transport infrastructure is poor. Long-rotation bush farming fallow systems, typical of this era of development, are widely regarded as 'sustainable'.

Economic development and population growth in coastal cities and lowland rural areas quickly change the upland economy. Commercialization (driven by the expansion of domestic and global markets), migration and natural population increase and the introduction of new technologies all create new pressures on the resource base. Customary law cannot easily accommodate such changes.[3] In this second phase of institutional development, traditional institutions governing resource use are swept aside (or merely recede);

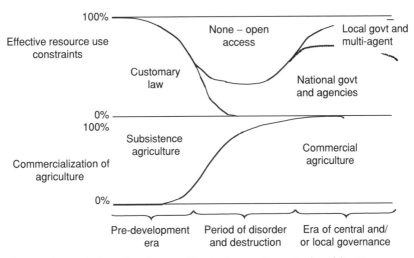

Fig. 1.1. Co-evolution of markets and institutions at the agricultural frontier.

the state assumes the lead role in controlling resource use and access, and new resource management institutions are imposed from above. But even though local governments and local offices of national resource management agencies may be established, these have no autonomy and little effective authority or capacity for enforcement. State power is thus low at the frontier; the resource base becomes, in effect, open access. Moreover, not only is there no political will for environmental or resource use measures that might reduce current income generation opportunities of people living in severe poverty amidst a

Table 1.1. Economic development and institutional evolution for environmental management.

	Subsistence economy	Early development	Late development
Economy and resource use	• Low population growth • Little or no migration • Resource abundance • Subsistence agriculture • Slash-and-burn • Long fallow periods	• High population growth • In-migration • Resource competition • Intensive agriculture • Commercial activity • Fallows shorten	• Population density falls • Out-migration • Resource use intensity falls • Mature agriculture • Commercial economy • Reforestation/protection
Institutional evolution	• Customary law • Community resource management	• State institutions with no practical checks • Property rights not well defined • 'Land grabs'	Ideally: • Central and local institutional change • Community stewardship • Property rights well defined Alternatively: • Local elites gain power • Incentives incompatible with benefits

perceived abundance of natural resources, but also the state typically *promotes* resource depletion as a means to generate household income and fiscal revenues. What follows is rapid deforestation, shortening of fallow periods and general degradation of soil, water and other environmental resources.

This 'period of disorder and destruction', as Burch (1990) describes it, was characteristic of many upland regions from the 1960s to the 1990s, during which time South-east Asia established itself as the world region with the highest annual deforestation rate (FAO, 2001), even as its largest resource-abundant economies (Indonesia, Malaysia and Thailand) expanded at historically unprecedented rates. Elsewhere, in highly repressed economies such as Myanmar, Laos and Vietnam, the state itself became the primary agent and entrepreneur of resource depletion, having closed off most other means to generate jobs, government revenue and foreign exchange.

In South-east Asia's more open economies, economic maturation and the end of the demographic transition has ushered in a third development phase, one in which lower labour force growth and rapid job creation outside of agriculture greatly reduce population-driven demand for upland resources, especially land. This has potentially large implications for the governance of such resources. In our view, this phase may result either in continued rapid resource depletion or in a shift to more conservative strategies. Which of these is more likely to occur? This will depend largely on the speed with which institutions catch up with the pace of economic growth. The countries of the region currently present a fascinating array of trends, with no clear, single pattern having emerged.

In this third phase, there is growing community demand for environmental quality and resource conservation. This trend is complemented by a more general decentralization of power and authority, as is currently taking place through formal means in all the large economies of the region. In some places, as local governments and communities assume control over some aspects of resource management, there may be a reassertion of customary laws, although in modified form. In the best situations, a combination of decentralization plus local demands for more ecologically friendly development is complemented by national laws and policies; in the best outcomes, national agencies, local governments and community groups collaborate to design (and more importantly, to implement) resource management and policies. In the worst cases, however, with continued growth of market demand for resources, reduced power at the national government level, and a 'business as usual' attitude on the part of local elites, 'disorder and destruction' continues, or even worsens.

A Review of Economy–Environment Linkages

Virtually all production generates some environmental damage in the form of pollution and/or natural resource depletion, and it follows that, other things being equal, such damage or depletion increases as an economy expands. It is well known, however, that the economy–environment relationship can be non-linear – and indeed, non-monotonic, a concept widely referred to as the

Environmental Kuznets Curve. Changes in economic structure occurring in the course of economic growth alter both the valuation and demand for environmental assets, and if sectors differ in their propensity to pollute or to use depletable resources, it follows that emissions and/or depletion rates will also change, a phenomenon that has been termed the *composition effect*.[4] The net environmental impacts of structural change can either be harmful or benign; more importantly, perhaps, the aggregate composition effect in reality has many elements, each with its own specific set of underlying economic and institutional determinants. For land and forest issues, this analytical approach enables research to move beyond trivially true assertions that deforestation and upland land degradation are the consequences of population pressure and 'market forces'. One thing that quickly becomes clear is that while *total* population growth in a country may justifiably be regarded as beyond the reach of policy except in the very long run, its spatial distribution, and the incentives that upland populations face when making resource use decisions, are very heavily conditioned by government policies, both those directly targeting such populations and activities, and also those operating at the broadest level of the economy.[5]

Upland agricultural development

The economies of Asia's uplands – usually defined officially by slope and/or altitude, but in practice referring also to relatively remote agricultural areas – differ both in structure and level of development by comparison with lowland zones. They are less densely populated and more dependent on agriculture and other resource-based industries; their populations are poorer, less healthy and less well educated. Market access is constrained by higher transport and transactions costs. Though accurate counts are impossible, the population living on 'fragile' lands in Asia and the Pacific is currently estimated at 469 million, or 25% of the total population (World Bank, 2003).[6]

As recently as a generation ago, many upland populations were effectively isolated from lowland and non-farm economies by infrastructural constraints, travel costs and even ethnic and political divisions. In spite of continued remoteness and relatively poor infrastructure, however, upland population trends, markets, policies and institutions are now very strongly influenced by the development of the overall economy. Roads and telecommunications networks integrate upland producers with the national economy. As markets expand, they create new economic opportunities, which upland and migrant populations are generally quick to seize. In so doing, they also alter the value of immovable resources such as forests and land. In a subsistence economy, such resources (and even labour) have values derived only from the requirements of the local economy, but market integration requires that resource valuations reflect returns obtainable in new uses.

Development policies have direct effects in South-east Asian uplands largely through infrastructure provision. The impact of road construction is huge, since it fractures natural landscapes, altering forest edges and interiors; at the same time it reduces transport costs and accelerates flows of migrants and information. As elsewhere in the developing world, road construction has a strong association with deforestation and the spread and intensification of agriculture (Cropper *et al.*, 1997; Andersen *et al.*, 2002).

The expansion of roads and markets, however, also conveys the *indirect* effects of policy distortions to upland resource use decisions, so economic and policy trends in industry and lowland agriculture become central to upland development (Coxhead and Jayasuriya, 2003), and national trends in food demand, agricultural technology and food policy can all have significant environmental consequences. Most obviously, agricultural support policies may stimulate the expansion of cultivated area at the expense of forest. The mechanisms for this change vary from country to country and over time, with contributions from state-sponsored land clearing for settlement programmes, commercial forestry and subsequent land conversion by corporate agribusiness enterprises, and deforestation and land clearing (as well as the intensification of bush fallow rotation systems) by 'subsistence' farmers (Angelsen, 1995). All land colonization, however, is driven by a combination of opportunity and necessity, and encouraged by the absence of well-defined and effectively enforced property rights over forest-covered land. As we have argued, the property rights problem itself is partly an artifact of government policies that identify forest-covered land (or land so designated, including cleared land above a certain slope or altitude) as a public resource, neither alienable nor disposable, without providing adequately for its protection from encroachment.

Lowland agricultural development

Whereas much of Asia was historically a region of food surplus and labour scarcity, 20th century population growth soon began to apply pressure on the agricultural land base. In the three decades after the Second World War, a period during which the region's population grew very rapidly, pressures on the land resource base began to climb, domestic food production *per capita* began to decline and the share of food in the value of imports began to rise. Investments in irrigation, and the introduction of yield-improving technology packages in the 1960s and 1970s, which centred on modern cereal varieties (the 'green revolution'), partially alleviated land scarcity by enabling production increases on existing land. In addition, states sponsored the colonization of new lands for food production through internal migration, supported by subsidized or publicly provided services such as land clearing and market and physical infrastructure.

Within agriculture as a whole, and in lowlands in particular, cereal production dominates land use and sectoral employment. It follows that technical progress in cereals – the green revolution – and food policies have significant effects on agricultural development. Governments of the net food-importing Asian economies have enshrined food security – or more strongly, self-sufficiency in cereals at the national or even subnational scale – as a basic plank of development policy (Barker and Herdt, 1985; David and Huang, 1996). The key instruments of self-sufficiency strategies have been quantitative restrictions on food trade, usually with monopoly control over imports assigned to a state agency. Historically, these policies, along with the industrialization policies discussed above, did much to determine resource allocation and investment flows both to agriculture as a whole, and to industries within that sector.[7] In the 1990s many such policies were converted to tariffs to comply with WTO rules. However, WTO rules on agricultural import policies by LDCs permit ample margins for maintenance of previous trade regimes, albeit with new instruments.

Outside of irrigated areas, the *direct* productivity impacts of infrastructural investments and new technologies in lowlands are generally small (David and Otsuka, 1994). When international trade is constrained, however, yield gains in lowland irrigated areas almost certainly diminish pressures for expansion of food production in uplands by driving down domestic grain prices. Rising labour productivity and labour demand in lowland agriculture also reduce incentives for labour migration to uplands (Coxhead and Jayasuriya, 1994; Hayami and Kikuchi, 2001; Shively, 2001). Historically, and even today, these indirect impacts of the green revolution confer environmental benefits in uplands, raising the opportunity cost of deforestation and land conversion (Coxhead and Jayasuriya, 2004; Shively and Fisher, 2004; Shively and Pagiola, 2004).

Philippine Economic Development and the Uplands

Since the 1950s, in response to new commercial opportunities, land used for agricultural purposes has expanded considerably in the Philippines. At the end of the Second World War, most sloping and high-altitude land in the country was still forested. Agricultural activities in these areas consisted primarily of maize, cassava and coffee production. Various forms of long fallow were employed by farmers. In the province of Benguet in Northern Luzon, commercial vegetable production intensified, and in the 1950s, farmers migrating out of this highland area introduced commercial cultivation of potatoes, cabbages and other temperate-climate vegetables to similar areas elsewhere. Commercial success in these crops, the introduction of new maize varieties, and the substitution of coffee and shrub crops with annual crops resulted in a steady intensification in upland land use in many areas of Mindanao, the country's largest and least densely populated island. Since the late 1970s, improvements in infrastructure, integration of frontier economies into national agricultural markets, and growing domestic demand for maize and temperate-climate vegetables have ensured that commercial agriculture in upland areas continues to expand.

The Manupali River Watershed

Physical characteristics of the watershed

The research highlighted in this book was conducted in the Manupali River watershed, located in the Philippine province of Bukidnon, over more than 10 years, beginning in 1992 (for the location of the study site, see Plate 1).

This watershed shares many of the historical and institutional features high-lighted in previous pages and consists of two political jurisdictions: the munic-ipality of Valencia, which covers the southern side of the upper watershed, and the municipality of Lantapan, which covers the northern side of the watershed. Lantapan covers the largest part of the watershed – 31,652 ha out of a total watershed area of approximately 38,000 ha (see Plate 2).

The municipality of Lantapan is bounded in the south by the left bank of the Manupali River, and in the north by Mt Kitanglad Range Natural Park, a major national park. From east to west the landscape begins as river flats devoted to irrigated lowland rice fields at 300–600 m above sea level, and extends through rolling areas planted to sugarcane and maize into a band of maize and coffee at 600–1100 m. Agriculture terminates in a mid-to-high alti-tude area of maize and vegetable production that extends from about 800 m into the buffer zone of the national park at 1500 m. Steeply sloped moun-tainsides extend to 2900 m. The watershed has a mean elevation of 1561 m and a highest point of 2939 m. More than two-fifths of the area has a slope of 40% or more; of the remaining area 29% is rolling and 27% is considered flat. An aerial photograph of the watershed is provided as Plate 3. A map of the watershed, including major towns, roads and waterways appears as Plate 4. Below our study site the Manupali River runs into a diversion dam feeding a 4000 ha gravity-fed irrigation system constructed by the Philippines' National Irrigation Authority in 1987. The entire system ultimately drains into the Pulangi River, one of the major waterways of Mindanao Island, about 50 km upstream from the Pulangi IV hydroelectric power generation facility, the largest such facility in Mindanao and one of the largest hydroelectric power–generating plants in the Philippines.

Mt Kitanglad Range Natural Park was established in 1996 and has a total area of 42,700 ha, with approximately two-thirds designated as a protected area and one-third designated as a buffer zone. Approximately 45% of the total park area is covered with old-growth forest and much of the park remains forested, in spite of adjacent human activity. As Garrity and Amoroso (1998) point out, the area is a rich repository of biodiversity and a source of many environmental services and forest products. Although Mt Kitanglad Range Natural Park is a relatively small ecosystem, it has been judged to be of high conservation value due to high rates of endemism of the vascular flora (Pipoly and Masdulid, 1995; Amoroso *et al.*, 1996). It has one of the great-est diversity of mammals and birds in the Philippines (Heaney, 1993) and an extremely high density of tree species. As an example, a 1-ha tree inventory conducted by Tabaranza (1995) and Pipoly and Masdulid (1996) indicated the presence of 43 species, 47% of them endemic to the Philippines. This is one of the highest known tree species density figures for any published tropical tree inventory.

Forest cover, however, is shrinking, and the buffer zone area surround-ing the park continues to degrade. GIS evidence from the watershed indicates substantial changes in land use and land cover over the past 20 years. A range of problems associated with this conversion of forest to agriculture in the watershed have been documented (Deutsch *et al.*, 1998). Findings from water monitoring in subwatersheds of the Manupali River show persistently rising total suspended solids (TSS) loadings, along with other indicators of diminution in watershed function. In some instances, events have been dramatic: in one recorded event, TSS loadings increased by 1000-fold within a 2-h period of heavy rain, to reach about 18 kg of soil in each cubic metre of water. The instability of some tributaries appears to be intensifying, as indicated by abrupt flooding and drought cycles resulting in washouts along main roads, property damage, and crop and soil losses. Such patterns of soil and water transport also pose health risks.

Water quality and human activity are closely linked in the Manupali River watershed. Abrupt increases in TSS have been observed after land conversion caused forest cover in a subwatershed to drop below 30% and the share of agricultural land to rise above 50%. Downstream, lowland farmers have become increasingly aware of the negative impacts of upland soil erosion due to heavy siltation of irrigation canals. In some cases, lowland water conveyance systems are only about 25% efficient. It also appears that the siltation problem in the downstream irrigation canal network has been intensifying. According to local observers, the amount of silt removed from downstream canals between 1993 and 1995 was more than twice the amount removed between 1990 and 1992. Last, but by no means the least, bacterial concentrations in tributaries of the Manupali River typically exceed WHO and US EPA safety standards for contact by a factor of 10–50.

The Manupali River is a source of water for drinking and irrigation, and – as mentioned above – is a tributary of the Pulangi River, which hosts several hydroelectric power plants, including the Pulangi IV facility. This power plant is a 255 MW run-of-the-river station that currently suffers from sedimentation so severe that its official expected life has been severely shortened.

Population and watershed economy

In many important respects – including a history of agricultural expansion, the persistence of poverty, and a worsening of on-site and downstream environmental quality – the Manupali River watershed shares features common to many tropical watersheds. The human population of the watershed is diverse. In the 1970s, Lantapan's population increased from 14,500 to 22,700 at an average annual rate of 4.6%; by 1994 the population exceeded 39,000 (Lantapan, 1994). Between 1980 and 1990 the population growth rate in this region averaged 4% per year, a rate that was much higher than the Philippine national average of 2.4%. In the most recent decade, however, population growth has fallen to roughly the same level as the national average.

The main indigenous population group is the *Talaandig*, who constituted 43% of the municipal population in 2001. Mt Kitanglad area is their ancestral home; they retain rights and have asserted claims over much of the land and resources in the watershed. Various migrant communities are also present, however, and immigration, mainly from the poorest provinces of the Central Philippines, has been a historically important determinant of activities.

The current population is a mix of indigenes with in-migrants of one, two or in a few cases, three generations' duration.

Not surprisingly, agriculture is the primary sector of employment and the major determinant of land use in the area. In 1988, 71% of provincial employment was in agriculture, 5% in industry and 23% in services; agriculture provides the primary income source for 68% of Bukidnon households. As is typical of a recently settled area, most Lantapan farms (about 70%, covering 80% of total farm area) were owned or in 'owner-like possession' in 1980, the last year for which agricultural census data are available. Farm sizes are small by upland standards: in Lantapan in 1980, the modal farm size class (1–2.99 ha) contained 46% of farms and 75% of all farms were smaller than 5 ha. Many households live close to the poverty line. In 1988, food accounted for 59% of household expenditures in Lantapan.

In 1973, approximately 28% of total land area in Lantapan was considered agricultural land and was used in the cultivation of temporary crops.[8] By 1994 this share had increased to nearly half of the watershed area. This expansion in land devoted to agriculture has largely involved the replacement of forest and permanent crops by annual crops. As is clearly seen in Table 1.2, over the 20-year period ending in 1994, the area of permanent forest shrank from approximately one-half to slightly more than one-quarter of the total land area. While a portion of this area was converted to shrubs or secondary forest, a much larger portion was changed to annual agricultural crops. Of special note is the expansion of land devoted to maize-vegetable systems; between 1973 and 1994, land allocated to these purposes increased from 17% to 33% of total land area. In early times, both logging and forest fires facilitated agricultural expansion. In recent decades, however, the profitability of commercial vegetable cultivation has been the primary impetus for forest encroachment, with decisive contributions from road development and the lack of well-defined and enforced property rights in land (Cairns, 1995). The expansion of vegetables and plantation crops in lieu of cereal crops in the area is also a result of favourable price and trade policies.

Table 1.2. Land use by slope (10% and greater), 1973 and 1994.

	Slope					
	10–20%		20–40%		40–90%	
Land use class	1973	1994	1973	1994	1973	1994
Dense forest	69.5	38.9	88.3	59.9	91.7	57.3
Shrub and tree (besides forest)	3.0	11.1	6.2	22.7	3.9	32.5
Shrub and tree (other distribution)	4.0	5.2	1.2	1.7	1.4	0.9
Agriculture	17.6	41.8	3.4	13.1	1.9	7.0
Grass	4.1	..	0.17	..	0.85	..
Bare soil	0.1	1.3	0.2	2.0	0.1	2.3
River and creek	1.7	1.7	0.5	0.5	0.1	0.1

Note: .. indicates data not available.
Source: Bin (1994), Tables 5.5 and 5.11. Data constructed from satellite imagery.

A number of economic phenomena have helped shape the observed pattern of land use, particularly the expansion of agriculture across the landscape and its intensification in certain locations. One of the most important economic drivers of these phenomena is movements in relative prices. On the commodity side, relative prices have changed in a discernible way over time, favouring annual crops such as maize and vegetables over perennial crops throughout most periods. Trends in input prices have also been influential in determining crop mixes, as the major crops differ considerably in their employment of factors. For these reasons, markets and national economic policies have played a key role in determining observed land use patterns in the watershed. For example, the profitability of maize and vegetable cultivation has been both directly and indirectly affected by Philippine government policies. Such policies consist mainly of market interventions intended to stabilize farm prices, trade interventions intended to defend the livelihoods of upland farmers and to reduce national dependence on imports, and technology interventions such as public support for research aimed at raising yields and reducing crop vulnerability to pests and diseases (Librero and Rola, 1994). The outcome of these interventions has been an increase in the area of maize planted in Bukidnon even as it has declined nationally in the Philippines (Coxhead, 1997, 2000).[9]

Important implications of expanded maize and vegetable production in the Manupali River watershed have been a steady decline in forest cover, an increase in erosion and sedimentation and an increase in the rate of pesticide use. Traditional crops, cropping systems and farming practices in the watershed remain problematic in this regard, although some technical solutions to the challenges of upland farming show promise. Furthermore, starting in 1998, at least ten commercial hog and poultry firms established a presence in the watershed; and in 1999 two banana companies were established in the middle and upper parts of the watershed. These activities place new strains on land and water resources and make it clear that steps to improve land use patterns in the watershed will require much more than just technical innovations. Given the rapid pace of decentralization that has taken place in the Philippines in the last decade, it may be possible for local institutions and local governments to play a facilitating role in altering patterns of agricultural and non-agricultural activity.

Overview of Chapters

The main goal of this book is to provide a focused analysis of the Manupali River watershed, including biophysical evidence on watershed degradation, economic and institutional causes and potential remedies. Findings are derived from research efforts based on data collected in the Manupali River watershed between 1992 and 2004. The chapters fall into three groups, defined roughly by their focus on: (i) the definition of challenges; (ii) the economic context of land use changes; and (iii) analyses conducted using the watershed as a unit of analysis.

Chapters 2 and 3 set the stage for subsequent chapters by examining patterns of agricultural development and environmental degradation in the watershed. In Chapter 2, Rola and Coxhead examine the history of agricultural development in the watershed and highlight institutional transitions that have taken place. The environmental implications of these economic and institutional changes are highlighted in Chapter 3, where Deutsch and Orprecio present 10 years of water quality data from the watershed. Chapter 3 establishes the link between changes in the physical characteristics of the landscape and changes in water quality, thereby providing the necessary broad perspective for the remainder of the book. Thematically, Chapters 4, 5 and 6 constitute a second section, and provide a set of closely related economic studies. In Chapter 4, Coxhead, Rola and Kim use time series data from local and regional agricultural markets to show how national markets have shaped incentives for agricultural expansion and resource allocation at the forest margin. In Chapter 5, Coxhead and Demeke then conduct an analysis at the farm level, using household panel data to examine the factors – including agricultural prices – associated with land use change. Having established the necessary evidence that land use in the watershed responds to changes in national prices, Chapter 6 presents analysis undertaken at a broader scale. We ask how the upland economy fits within the overall Philippine economy, using results from a computable general equilibrium model. Taken together, the results from this second group of chapters clearly establish the economic context in which land use change has occurred in the Manupali River watershed.

Chapters 7–11 constitute a third thematic section of the book and share two features: all take the watershed as their unit of analysis and all focus on possible remedies for landscape scale resource degradation. In Chapter 7, Ella reports results from simulations of soil erosion and sediment yields in subwatersheds of the Manupali River, asking how changes in land use might alter rates of erosion and downstream sediment yields. In Chapter 8, Paningbatan uses GIS methods to map a series of erosion 'hotspots' in the watershed. This approach shifts the focus towards predictive analysis, in the process highlighting the potential importance of targeting efforts to improve land use practices. Exactly what might constitute agronomically feasible alternatives to traditional annual crop agriculture is the subject of Chapter 9, in which Midmore and colleagues assess a range of alternative farming systems for the uplands and their likely consequences for the agricultural landscape.

Two subsequent chapters return the book to an economic focus. In Chapter 10, Shively and Zelek address the question of how changes in policies, especially economic policies, might influence land use decisions. Using a simulation model they investigate combinations of local and national economic policy changes designed to alter the incentives for various forms of crop agriculture in the watershed. In addition to examining the impact of a range of policies on household welfare and downstream environmental outcomes, they assess the distributional and budgetary impacts of various policies, asking what combinations of policies might achieve environmental goals in the most cost-effective manner. In Chapter 11, Pagiola, delos Angeles and Shively highlight the potential role that might be played by providing payments for environmental services

(PES) to achieve environmental improvements. The book concludes with Chapter 12, in which Coxhead, Rola and Shively explore the evolving interface between national development policies and local institutions as they shape the upland environment of the Manupali River watershed.

Notes

[1] Concerns that erosion from upland farming creates sediment in downstream irrigation systems, reducing both their productivity and expected life, are especially important in light of evidence that a lack of reliable water supply precludes expansion and intensification of agriculture in many low-income areas of Asia (Myers, 1988; Svendson and Rosegrant, 1994; Rosegrant *et al.*, 2001). As an example, estimates from the Philippines suggest 74–81 million t of soil is lost annually, and 63–77% of the country's total land area is affected by erosion (Forest Management Bureau, 1998). Sedimentation has reduced storage capacity at all of the Philippines' major reservoirs, and has measurably affected domestic water consumption, power generation and irrigation. Over the last 25 years dry season irrigated area has fallen by 20–30% in several of the country's key irrigation systems (Forest Management Bureau, 1998).

[2] Parts of this section draw on Rola and Coxhead (2005).

[3] For an example of how customary rules of resource access are sometimes unable to adapt to rapid changes in technology, see Shively (1997).

[4] The composition effect is one of three normally identified in analyses of growth-environment linkages. The other two are the *scale effect* (the additional demand on environment and natural resources due to economic expansion), and the *technique effect*, capturing secular changes in technology and non-homothetic preferences, due to changes in the capital stock and changes in incomes, respectively. Unlike the composition effect, the signs of both the scale and technique effects are hypothesized to be unambiguous: scale effects increase environmental damages and technique effects lessen them. See Antweiler *et al.* (2001).

[5] See Binswanger (1991) for an early analysis of policy influences on deforestation in the Brazilian Amazon.

[6] 'Fragility' is defined by criteria relating to aridity, slope, forest cover and soil type.

[7] South-east Asia's two largest food importers, the Philippines and Indonesia, each achieved self-sufficiency in rice (in 1970 and 1985, respectively) as the result of huge investments of public funds in rice production, R&D, and related infrastructure, with substantial land and other resources diverted from production of other crops (Barker and Herdt, 1985).

[8] This area excludes areas devoted to production of coffee, rubber, abaca and other tree and shrub crops.

[9] In Chapter 6 we use an economy-wide model of the Philippines to show that, at constant prices, even modest technical progress in maize production (which has the same effects on farm profitability as a price rise) increases the area planted to maize.

References

Amoroso, V.B., Acma, F. and Pava, H. (1996) Diversity, status, and ecology of Pteridophytes in three forests in Mindanao. In: Camus, J.M. and John, R.J. (eds) *Pteridology in Perspective*. Royal Botanic Gardens, Kew, UK, pp. 53–60.

Andersen, L.E., Granger, C.W.J., Reis, E.J., Weinhold, D. and Wunder, S. (2002) *The Dynamics of Deforestation and Economic Growth in the Brazilian Amazon*. Cambridge University Press, Cambridge.

Angelsen, A. (1995) Shifting cultivation and 'deforestation': a study from Indonesia. *World Development* 23, 1713–1729.

Antweiler, W., Copeland, B.R. and Taylor, M.S. (2001) Is free trade good for the environment? *American Economic Review* 91, 877–908.

Asian Productivity Organization (1995) *Soil Conservation and Watershed Protection in Asia and the Pacific*. Asian Productivity Organization, Tokyo.

Barker R. and Herdt, R.W. (1985) *The Rice Economy of Asia*. Resources for the Future, Washington, DC.

Bin, L. (1994) The impact assessment of land use change in the watershed area using remote sensing and GIS: a case study of the Manupali watershed, the Philippines. Master's degree thesis. School of Environment, Resources and Development, Asian Institute of Technology, Bangkok, Thailand.

Binswanger, H.P. (1991) Brazilian policies that encourage deforestation in the Amazon. *World Development* 19, 821–829.

Burch, W.R. (1990) Foreword. In: Poffenberger, M. (ed.) *Keepers of the Forest. Land Management Alternatives in Southeast Asia*. Ateneo de Manila University Press, Quezon City, Philippines.

Cairns, M. (1995) Ancestral domain and national park protection: mutually supportive paradigms? A case study of the Mt. Kitanglad Range National Park, Bukidnon, Philippines. Paper presented at a workshop on Buffer Zone Management and Agroforestry, Central Mindanao University, Musuan, Philippines, August. Mimeo.

Coxhead, I. (1997) Induced innovation and land degradation in developing country agriculture. *Australian Journal of Agricultural and Resource Economics* 41, 305–332.

Coxhead, I. (2000) Consequences of a food security strategy for economic welfare, income distribution, and land degradation. *World Development* 28, 111–128.

Coxhead, I. and Jayasuriya, S. (1994) Technical change in agriculture and the rate of land

degradation in developing countries: a general equilibrium analysis. *Land Economics* 70, 20–37.

Coxhead, I. and Jayasuriya, S. (2003) *The Open Economy and the Environment: Development, Trade and Resources in Asia*. Edward Elgar, Cheltenham, UK.

Coxhead, I. and Jayasuriya, S. (2004) Development strategy, poverty, and deforestation in the Philippines. *Environment and Development Economics* 9(5), 613–644.

Cropper, M., Griffiths, C. and Mani, M. (1997) Roads, population pressures and deforestation in Thailand, 1976–1989. Policy Research Working Papers. The World Bank, Washington, DC.

David, C.C. and Huang, J. (1996) Political economy of rice price protection in Asia. *Economic Development and Cultural Change* 44, 463–483.

David, C.C. and Otsuka, K. (eds) (1994) *Modern Rice Technology and Income Distribution in Asia*. Lynne Reinner Publishers, Boulder, Colorado, and London; and International Rice Research Institute, Baños, Philippines.

Deutsch, W.G., Busby, A.L., Oprecio, J.L., Bago, J.P. and Cequia, E.Y. (1998) Community-based water quality indicators and public policy in rural Philippines. In: *Economic Growth and Natural Resource Management: Are They Compatible?* Proceedings of the SANREM CRSP Philippines 1998 Annual Conference, Pines Hills Hotel, Malaybalay, Bukidnon, Philippines, 18–20 May 1998.

Dixon, J.A. and Easter, K.W. (1986) Integrated watershed management: an approach to resource management. In: Easter, K.W., Dixon, J.A. and Hufschmidt, M.M. (eds) *Watershed Resources Management*. Westview Press, Boulder, Colorado.

Doolette, J.B. and Magrath, W.B. (1990) Strategic issues in watershed development. In: Doolette, J.B. and Magrath, W.B. (eds) *Watershed Development in Asia*. World Bank Technical Paper No. 127. The World Bank, Washington, DC.

Easter, K.W., Dixon, J.A., and Hufschmidt, M.M. (eds) (1986) *Watershed Resources*

Management. Westview Press, Boulder, Colorado.

FAO (Food and Agriculture Organization of the United Nations) (2001) *Global Forest Resources Assessment 2000*. FAO, Rome. Accessed 20 August 2004 at www.fao.org/forestry/fo/fra/main/pdf/main_report.zip

Forest Management Bureau (1998) *The Philippines Strategy for Improved Watershed Resources Management*. Philippines Department of Environment and Natural Resources, Manila.

Garrity, D.P. and Amoroso, V.B. (1998) Conserving tropical biodiversity through local initiative. In: *Economic Growth and Natural Resource Management: Are They Compatible?* Proceedings of the SANREM CRSP Philippines 1998 Annual Conference, Pines Hills Hotel, Malaybalay, Bukidnon, Philippines, 18–20 May 1998.

Hayami, Y. and Kikuchi, M. (2001) *A Rice Village Saga: Three Decades of Green Revolution in a Philippine Village*. International Rice Research Institute, Los Baños, Philippines.

Heaney, L. (1993) Survey of vertebrate diversity in Mt. Kitanglad Nature Park. Unpublished Manuscript. Philippine National Museum, Manila.

Heathcote, I. (1998) *Integrated Watershed Management: Principles and Practice*. John Wiley & Sons, New York.

Lantapan, Municipality of (1994) Municipal agricultural and demographic database

Librero, A.R. and Rola, A.C. (1994) Vegetable economics in the Philippines. Paper presented at the AVRDC Workshop on Agricultural Economics Research on Vegetable Production Systems and Consumption Patterns in Asia, Bangkok, Thailand, 11–13 October 1994.

Munasinghe, M. (1992) *Water Supply and Environmental Management*. Westview Press, Boulder, Colorado.

Myers, N. (1988) *Natural Resource Systems and Human Exploitation Systems: Physiobiotic and Ecological Linkages*. The World Bank, Washington, DC.

Naiman, R.J. (1995) *Watershed Management: Balancing Sustainability and Environmental Change*. Springer-Verlag, New York.

Perrings, C. (1989) An optimal path to extinction? Poverty and resource degradation in an open agrarian economy. *Journal of Development Economics* 30, 1–24.

Pipoly, J. and Masdulid, D. (1995) The vegetation of a Philippine submontane forest, Kitanglad Range. Unpublished, cited in Garrity *et al.* (1998).

Pipoly, J. and Masdulid, D. (1996) Tree inventory in submontane forest of Mt. Kitanglad. *Proceedings of the Flora Malesiane Symposium*. Royal Botanic Garden, Kew, UK.

Rola, A.C. and Coxhead, I. (2005) Economic development and environmental policy in the uplands of Southeast Asia: challenges for policy and institutional development. In: Colman, D. and Vink, N. (eds) *Reshaping Agriculture's Contribution to Society: Proceedings of the Twenty-Fifth International Conference of Agricultural Economists*. Blackwell, Malden, Massachussetts, and Oxford, UK, 243–256.

Rosegrant, M.W., Paisner, M.S., Meijer, S. and Witcover, J. (2001) *Global Food Projections to 2020: Emerging Trends and Alternative Futures*. International Food Policy Research Institute, Washington, DC.

Shively, G.E. (1997) Poverty, technology, and wildlife hunting in Palawan. *Environmental Conservation,* 24, 57–63.

Shively, G. (2001) Agricultural change, rural labor markets, and forest clearing: an illustrative case from the Philippines. *Land Economics* 77, 268–284.

Shively, G. and Fisher, M. (2004) Smallholder labor and deforestation: a systems approach. *American Journal of Agricultural Economcs* 86, 1361–1366.

Shively, G. and Pagiola, S. (2004) Agricultural intensification, local labor markets, and deforestation in the Philippines. *Environment and Development Economics* 9, 241–266.

Svendson, M. and Rosegrant, M. (1994) Irrigation development in Southeast Asia beyond 2000: will the future be like the past? *Water International* 19, 25–35.

Tabaranza, B. (1995) Fauna survey in Songko, Lantapan. Narrative report submitted to SANREM CRSP Project Office, Malaybalay, Philippines.

World Bank (1992) *World Development Report 1992: Development and the Environment*. The World Bank, Washington, DC.

World Bank (2003) *World Development Report 2003: Sustainable Development in a Dynamic World*. The World Bank, Washington, DC.

2 Agricultural Development and Institutional Transitions*

A. C. ROLA[1] AND I. COXHEAD[2]

[1]University of the Philippines at Los Baños, College, Laguna 4031, Philippines, e-mail: arola@laguna.net; [2]Department of Agricultural and Applied Economics, University of Wisconsin-Madison, 427 Lorch Street, Madison, WI 53706, USA; e-mail: coxhead@wisc.edu

Introduction

Economic development in the uplands is subject to tremendous influence from events in other sectors and regions, as subsequent chapters in this book will make clear. Through markets and migration, policies directed at specific 'lowland' sectors can also affect upland resource valuations, patterns of land use and production, and thus environmental outcomes (Coxhead and Jayasuriya, 2003). The ways in which these external influences affect uplands and forests, however, depend critically on the structure of the upland economy and in particular on the legal framework and other institutions that constrain or encourage resource allocation and exploitation decisions.

It was argued in Chapter 1 that economic development in the uplands has been accompanied by distinct phases of institutional development. These phases were highlighted in Fig. 1.1. The early phase is defined by the erosion of customary law, and later phases by the assertion of authority by the central state – often after a hiatus described by Burch (1990) as a 'period of destruction and disorder'. In some localities we now see a third set of institutional arrangements on the rise, as local governments and communities assume control over some aspects of resource management. In some instances, there is a move to reassert customary laws, albeit in modified form.

This chapter addresses institutional evolution with particular reference to South-east Asia and the Philippine case, and focuses on those institutions most critical to forests and upland land resources. The discussion addresses resource management policies as well as access and control, that is, the property rights attached to the resource in question, as primary factors conditioning household and community decisions on natural resource management. In the Philippine

---

*Parts of this chapter are drawn from Rola and Coxhead (2005) and are used with permission.

case study we identify some examples of recent institutional innovations that appear to satisfy many requirements for sustainable decentralized natural resource management in the long run.

Property Rights over Land and Forest

The centralization of government in general, and of control over natural resource assets in particular, has long been a feature of governance in developing countries. In the Philippines, Spanish-era law asserted the State's ownership of all land unless a decree was issued to the contrary, and this precedent has 'remained the theoretical bedrock upon which Philippine national laws were based...land not covered by official documentation is considered part of the public domain...regardless of how long [it] has been continuously occupied and cultivated' (Lynch, 1987: 270). This doctrine persisted through the American administration and independence. Although by the mid-1980s the area of declared public land had shrunk to only 62% of the total, it still covered 90% of uplands (Lynch, 1987: 270). A 1975 presidential decree explicitly prevented the occupants of uplands from acquiring private property rights, at the same time as it declared that existing occupants were immune from prosecution (Lynch, 1987: 284). Given the lack of documentation and difficulty of enforcement, this act effectively legislated open access to forestlands by individuals.

In Indonesia, the Dutch colonial regime was in general content to rule through local leaders. However, the colonizers asserted ownership over land and forest once the value of those assets had been raised by trade in the mid–late 19th century. Even in the independence era, Dutch colonial practice persisted in the Basic Forestry Law of 1967, which designated 74% of total area (and 90% of the Outer Islands) as 'state forestland' under the control of the Ministry of Forestry, nullifying traditional law or *adat*. As Fay and Surait (2002: 126) describe it: '[T]he creation of the state forest zone nullified local customary rights, making thousands of communities invisible to the forest management process and squatters on their ancestral land.' Other post-colonial regimes in South-east Asia, as well as in Thailand, passed similar laws around this time.

Though the primary incentive for control over land and natural resources was its exploitation for economic growth, motivation for continued central ownership and control was as much political as economic or managerial:

> The new independent nation-states that arose following the Second World War have shown little interest in revitalizing local-level systems of authority . . . (they) do not relish the thought of local political forces that might challenge the legitimacy and authority of the national government. This means that natural resources have become the 'property' of the national governments in acts of outright expropriation when viewed from the perspective of the residents of millions of villages. This expropriation is all the more damaging when national governments lack the rudiments of a natural resource management capability
>
> (Bromley, 1991: 127)

National planners regarded natural forests as resources to be exploited for national development. The harvesting and exporting of timber and forest products was a means to finance modern agricultural and economic growth (see ADB, 1969). In the Philippines, for instance, veneer, logs and lumber were among the top export earners from before the Second World War until the early 1980s. Indonesia, likewise, depended heavily on foreign exchange earned through the exploitation of its forests prior to the oil booms and industrial growth of the 1970s and 1980s (Resosudarmo, 2002).

Forest Management

The question of who governed the forests during this transitional period of development is more complex than the applicable laws indicate. While most forest resources in South-east Asia were claimed by the state and governed at the national level (Table 2.1), in reality these resources were controlled by a number of actors, with or without the blessing of national government. In several countries (and regions within countries), alliances and conflicts among different national agencies, the military, local elites, and domestic and foreign timber corporations determined forest access. In others, local communities successfully maintained at least some degree of influence, usually due to difficulty of access. Tribal and cultural groups retained much effective control over forest resources in the highlands of Vietnam, Laos and Thailand, for example, in spite of highly centralized forest management policies (Poffenberger, 2000).

In the Philippines and Indonesia, however, central government and its private sector allies have been more dominant (Kummer, 1992). Alliances (formal and legal, or otherwise) between timber corporations and the state (as nominal owner of the forests) undoubtedly contributed to very high rates of forest clearing. In the Philippines, for example, the Marcos administration (1965–1986) and private companies forged a Timber License Agreement (TLA) which permitted corporate entities to cut trees in a forest area of not more than 100,000 ha for a period of 25 years, with no effective responsibilities for forest management, replanting or environmental controls. This policy and its successors are partly responsible for the Philippines having recorded the highest annual deforestation rate among South-east Asian countries in 1980–1990 (3.3%), and in 1990–2000 (1.4%) (FAO, 2001).

Policies influencing forest and land management have also arisen as side effects of other facets of particular countries' development strategies. Most well-known among these are the various internal migration initiatives of the 1960s to 1980s, in which government agencies cleared and developed virgin land at the frontier to house and sustain sponsored migrants. Under these programmes large areas, most notably in Malaysia and Indonesia, were converted to plantations and upland fields, and usually supplied with irrigation and other infrastructure at considerable public expense. However, the degree of control over land use exerted by government agencies in transmigration areas varied. In Malaysia, where the main focus was on development of rubber and other plantation crops, federal land development agencies maintained a high degree

Table 2.1. Policy context for forest management, selected Asian countries.

Policy Period	Country				
	Cambodia	Indonesia	Vietnam	Philippines	Thailand
1970s to 1990s	• Military, local elite and foreign corporations control the forests • All forestland belongs to the state	• Centralized control • Overdependence on forest resources for national income • Log ban and vertical integration of the forest industry	• Government nationalized large areas of land in the 1950s and early 1960s • Local residents have no access to forest lands	• Centralized governance of natural resources • Natural resource extraction is a primary vehicle for development	• Cultural communities govern forest management • National Forest policy/Land reform act was implemented but did not specify environmental rights and responsibilities to communities
1990s onwards	• Community involvement is encouraged in forest management • Comprehensive policy framework to clarify tenure right and responsibilities needed; move to decentralization	• Community-based forest management • Timber plantation management	• Private households replace state forest enterprises • 1993 Land Law gave local inhabitants extensive use rights over agriculture and forestlands	• IP rights recognized • Logging ban • Decentralization of some DENR function to local government • Community-based forest management as a national strategy	• National parks and sanctuaries were established but in conflict with policy of promoting community forest management • No formal legal basis for community resource management

Sources of basic data: Poffenberger (1990, 2000); Nilo (2000); Resosudarmo (2002).

of land use control, whereas in Indonesia, where national agencies had little effective authority, migrant communities frequently adapted infrastructural and land facilities to their own uses (Gérard and Ruf, 2001).

By the late 1980s, as commercially exploitable stocks dwindled, damaging side-effects of deforestation became more readily apparent, domestic and foreign conservation movements gained voice and forest policies began to change. Thailand and the Philippines both imposed bans on commercial logging and on the export of raw timber around 1990.[1] Like the resource exploitation policies that preceded them, however, these conservation measures also assumed an often unrealistic level of central control. Timber harvest restrictions were difficult to implement in the absence of appropriate incentives for loggers and exporters, and were widely circumvented. Efforts to provide incentives for sustainable forest management by corporations, primarily through long-term forest leases, met with only limited success.[2]

The Shift to Local Ownership and Management

The global trend towards decentralized government since the 1990s has begun to transform forest and land management in Asia, just as it has many other areas of policy and administration. In most countries, however, the evolution of forest policy away from centralized exploitation (and more recently, from centralized attempts at conservation) towards one in which communities and their representatives occupy centre stage is very new. The recognition of land or forest *ownership* by upland communities is even more novel – in spite of earlier legal steps in the same direction – and remains quite incomplete.

Many governments of South-east Asia subscribed to the ideal of local resource management in the 1990s. This shift usually coincided with broader programmes of political decentralization. Community-based forest management (CBFM) programmes, in which some resource decisions are deconcentrated or devolved to local jurisdictions (albeit within an overarching programme of central control) became widespread. The period of decentralization also saw some governments commit to ceding broader legal powers, including ownership. But the shift has been incomplete, and the legal basis for devolved management and ownership remains weak. In Cambodia, for instance, community involvement in forest management is encouraged, but the absence of documented forest rights and responsibilities leaves communities with no legal authority (Poffenberger, 2000). Similarly, Thailand has implemented a National Forest Policy which includes decentralized management powers, but does not specify the environmental rights and responsibilities of communities. In Vietnam, the 1993 Land Law conferred land use rights to communities and individuals, but the ownership of land and forests still rests with 'the people', that is, the state (Tachibana *et al.*, 2001). In part for these reasons, devolution has not provided a complete set of solutions for forest degradation and depletion. Policy conflicts have continued, fuelled both by the economic pressures of growth and by mismatches between central and local powers (see Chapter 1). National agencies continue to make and implement

policy, for example declaring protected areas or initiating reforestation programmes, without local consultation. In other cases, local communities conduct their affairs without reference to national plans or regulations. These are
most obviously seen in post-Suharto Indonesia, where local administrations,
empowered by several large legal steps towards decentralization in 2000, have
awarded themselves increased powers both *de jure* and *de facto,* gambling
that central authorities will not institute effective counter measures (Colfer and
Resosudarmo, 2002).

The Philippines has made considerable progress towards forest and upland
tenure reforms for environmental management, especially since the passage of
the 1991 Local Government Code. Implementation of the Integrated Social
Forestry Program (part of the Comprehensive Agrarian Reform Program
passed in 1987 and covering all agricultural lands, including public alienable
and disposable lands) has facilitated granting of tenure to forest occupants. In
many upland areas, local inhabitants can secure a Certificate of Stewardship
Contract (CSC) giving them exclusive use and occupancy rights to public
forestlands for a period of 25 years with possible renewal. Individuals, families
and local communities enter into a contract with the government for this purpose. But it is the duty of CSC holders to engage in soil conservation, sup-
pression of forest fires, and conservation of forest growth in their areas of
responsibility (Magno, 2003). Other laws passed in the 1990s also devolved
forest management, among them the 1997 Indigenous People's Rights Act
(IPRA), which purports to recognize land claims based on ancestral domain.
But ambiguities in the law persist, and its implementation has not been
smooth. Systemic political and institutional weaknesses mean that though new
legal arrangements are introduced, older ones do not go away, creating layers
of conflicting authority. It is not uncommon for putatively public land in the
Philippine uplands to be the subject of multiple ownerships claims, each
appealing to a different piece of legislation.

Policies not directly linked to land or forest tenure have also affected
upland land use and development. Some programmes, such as internal
migration schemes, have no direct relation to natural resource management
but nevertheless exert indirect influences by creating incentives or opportunities to change land use. Other measures directly target particular aspects
of the upland economy and upland environments. These include extension,
research and development aimed at generating new or more efficient upland
farming techniques, forest protection and reforestation measures, and soil
conservation projects. Agroforestry has also become a widely promoted
technology package aimed at upland protection and rehabilitation. While
official resource management strategies have by and large relied on these
direct approaches, however, farmers have typically been reluctant adopters.
Tenure insecurity and the broader economic environment are both influential
constraints; so too are the often contradictory signals sent indirectly by land
development and resettlement policies, which encourage agricultural expansion and intensified land use, usually with little regard to environmental consequences (Rola and Coxhead, 2002). Contradictory policies persist because
they are the domains of different agencies (for example the Agriculture or

Environment ministries) with no overarching coordinating mechanism (e.g. TDRI, 1994).

In summary, policies that support community-based stewardship of natural resources are evolving slowly in South-east Asia, after a long period of more or less open access. But the structures that are emerging are quite different from earlier customary governance systems, not least in that there is frequently an explicit commitment to collaboration between the state (national or local governments) and civil society. Incomplete decentralization of other areas of administrative control may have complicated the forest and land management process. In many cases, local control does not imply governance by the community; rather, there is an ongoing evolution of local power that mimics the centralized system it has replaced. Local elites, foreign interests and other actors now also have access to resources – sometimes with even less oversight than before. Whether decentralization is a better arrangement than centralized control of resources, therefore, remains to be seen.

Decentralization and Environmental Management

Decentralization, like globalization, undermines the potency of the traditional resource management model based on regulatory constraints designed and implemented by central agencies. After years of failed attempts at centralized management, opinion in the 1990s turned decisively in favour of local approaches. In South-east Asia, as noted above, this shift coincided with the decentralization of numerous other government functions. Can local governments do a better job of resource management than central governments? Some advantages are clear. For example, local administrations often have specific knowledge of local environmental and economic conditions, and therefore should be in a good position to fine-tune policy. But disadvantages also surface. Four of these are especially widespread: externalities, lack of accountability, overlapping mandates and capacity constraints.

Externalities. Jurisdictional boundaries do not typically coincide with relevant natural resource boundaries (such as watersheds), leading to problems of horizontally overlapping control areas and unresolved externalities. When the correspondence between the boundaries of political jurisdictions and the optimal units for natural resource management is inexact, environmental policy managers have an incentive to overexploit the resource. In South-east Asia this problem is most clearly visible in the management of watersheds and river basins, and is compounded when economic and population growth increase local demands on water and land resources. In Vietnam, for instance, conflicts have emerged between upstream coffee irrigators and downstream rice producers, and in China, development activities in upstream locations lead to social conflict with neighbouring villages over pollution, erosion and siltation in downstream locations (Dupar and Badenoch, 2002). These are by no means exceptions; similar tensions have been documented throughout the region. Although the textbook solution is easy – recentralize control to internalize external costs – in practice it proves difficult to 'take back' devolved powers, as the Indonesian government has discovered while trying to mend some of the more egregious errors in the 2000 decentralization laws.

Lack of Accountability. Accountability of local administrations is a critical constraint to socially beneficial local decision making. At the macro level, accountability requires institutional checks and balances on the actions of local governments, private businesses and even NGOs. A strong external audit system is considered to be critical in ensuring macro-level accountability (Manasan *et al.*, 1999). Local governments could be more fiscally responsible and accountable if they were endowed with greater power to tax, because they are closer to the constituents that they tax. Currently, local governments in most countries in the region are authorized to collect real property and local business taxes only (Gonzales and Mendoza, 2002). In all South-east Asian countries, a large portion of local government income is still supplied by central government. When local funds are limited, fiscal incentives exist for local governments to promote growth of polluting industries and accelerate resource extraction (Rola *et al.*, 2003). This has also been in evidence in Indonesia since the fall of the New Order regime (World Bank, 2000a; Lewis, 2001).

At the local level, accountability is frequently determined by the availability of constitutional and practical instruments by which communities acquire 'voice' in the formation and implementation of local policy. While decentralization 'does not guarantee that local communities will reap more benefits and be more interested in sustainable environmental management, it does increase the chances that this will happen' (Manasan, 2002). As Table 2.1 illustrates, participatory approaches to natural resource management, that is, those involving the community directly in addition to local government, improve the likelihood that the latter will be held accountable for resource management decisions.

Overlapping Mandates. A third and related problem arises from incomplete and uneven decentralization of functions. This often means that the mandate assigned to a local agency may not be matched by the authority vested in it. One of the most common problems reported in the literature is the inadequate coordination between line agencies and the elected local authorities that have assumed the management of their environmental resources. Decentralization laws in the Philippines, Thailand and Indonesia have failed to provide a clear division of responsibilities between local government and national line agencies, for example by increasing budgetary transfers without allocating new expenditure responsibilities (World Bank, 2000a,b). In all these countries, national line agencies operating at a local level tend to remain very much committed to national programmes, even if these are inconsistent with local goals. In the case of conflict, the policies of national governments tend to override local preferences, sometimes precipitating vehement and even violent confrontations of communities and representatives of the state.

Uneven decentralization also means that overlapping mandates among central agencies may cause policies applied by one agency to cancel the effects of policies applied by others (Manasan, 2002; TDRI, 1994). Hence, one challenge of decentralization is to 'embed efforts in a framework that promotes

Table 2.2. Decentralization dynamics: selected cases in South-east Asia, 2002.

Study site	Decentralization dynamics		Remarks
	Actors and powers	Outcomes	
Nghe An province, Vietnam	• District forest agencies gain new responsibilities	• Lack of genuine consultative process • Detailed technical specifications from centre create adverse effects for livelihood and environment	• Lack of accountability • Mismatched mandate and authority
Ratanakiri province, Cambodia	• Budgets provided by the national government from donor funds • Provincial officials provide support to legitimize local initiative and interests.	• Local people lack rights to defend resources from external commercial interests, thus undermining environmental, livelihood benefits of decentralization	• Accountability, mismatched mandate and authority
Luang Phabang province, Laos	• Villages have increased role in land use planning • Provincial officials have increased roles in regional development planning.	• Local consultative process leads to collective action • Line agencies lack technical expertise and support to carry out their duties	• More accountability, but mismatched mandate and authority • Capability constraints
Chiang Mai province (Mae Chaem watershed), Thailand	• Villages have increased voice in development decision making through elected tambon committee	• Lack of clear division of labour between local government and line agencies leads to lapses, tensions in natural resource management • Existing community-based organizations and networks strengthen decentralization process	• Mismatched mandate and authority

continued

Table 2.2. (*continued.*) Decentralization dynamics: selected cases in South-east Asia, 2002.

Study site	Decentralization dynamics		
	Actors and powers	Outcomes	Remarks
Bukidnon province (Manupali Watershed), Philippines	• Local government has fiscal powers	• Fiscal decentralization is an avenue to exploit natural resources to generate funds. This can accelerate watershed degradation	• Lack of accountability
	• Community-based water watchers have no legal support	• Watershed level coordination among different LGUs is lacking	• Externality problems

Source: Except for the Philippines, data were taken from Dupar and Badenoch (2002). The Philippine case study is from Sumbalan (2001), Rola *et al.* (2003) and Rola *et al.* (2004). Remarks column are the authors' interpretations.

overall national goals of economic and administrative integration, environmental quality and revenue generation while allowing sufficient flexibility in local implementation to meet unique local conditions' (Dupar and Badenoch, 2002). *Capacity Constraints.* Finally, local governments face capacity constraints in the conduct of analysis, policy formation, and fiscal powers needed to implement resource management measures (Coxhead, 2002). In most countries, environmental databases are weak, which undermines both national and local policy capability (World Bank, 1995). Local officials usually lack technical capacity for policy design and implementation, and for project monitoring and evaluation. In the Philippines, most technical experts for environmental management report to the central office; only a very lean force is assigned at the field level (Manasan, 2002).

Decentralization and Upland Resource Management Institutions in the Philippines

In the remainder of this chapter we survey recent institutional trends in the Philippines, focusing on the local government of Lantapan. The survey highlights challenges to decentralized management of the types just identified, but also indicates some recent innovations that carry the seeds of improved institutions and management strategies for the future.

Since passage of the Local Government Code in 1991, environmental management mandates and policy powers are distributed across all levels of governance in the Philippines, funding and other forms of administrative support come from Congress and from local government. Technical support in the preparation of provincial and municipal plans comes from line agencies and local government units, with assistance from the academe and from development projects. In a new departure, responsibility for much upland resource policy is increasingly in the hands of now-evolving multi-level, multi-sectoral institutions (see below).

The national mandate is abstract: Philippine law states that it is the duty of the national government to maintain ecological balance. The Department of the Environment and Natural Resources (DENR) is tasked to lead in this function, although its mandate naturally overlaps with those of the Department of Agriculture, the National Irrigation Administration and several other bodies. There has, however, been a serious effort to create an overarching coordination mechanism. Following the Earth Summit in 1992, the national government created the Philippine Council for Sustainable Development (PCSD), with the head of the National Economic and Development Authority as its chair. The PCSD is mandated to oversee and monitor the implementation of the Philippine Agenda 21 (PA21), the Philippines' blueprint for sustainable development, by providing the coordinating and the monitoring mechanisms for its implementation. Like its global namesake, PA21 as an expression of the national mandate to 'protect the environment' has set out broad goals and strategies, that is, involvement of non-governmental organizations (NGOs) and peoples organizations (POs), people empowerment, etc., but doesn't dwell much on specifics.

The 1992 Local Government Code (LGC) initiated a period of devolution of national mandate, including some of the DENR. The present day is thus a transition period in which national agencies are slowly devolving responsibilities to local governments; but the DENR remains the least decentralized of all Philippine government agencies, with only 4% of its staff and 9% of its budget located outside Manila, in contrast with averages of 51% and 12%, respectively for all agencies (Manasan, 2002).

At the sub-national level, much of the devolved natural resource management responsibility under the LGC bypasses provinces, moving power directly to municipalities. In practice, however, provincial governors and their administrations retain considerable influence over local decision making. Part of this takes place formally, through the exercise of their supervisory and coordination functions over municipalities and in particular through the Provincial Environment and Natural Resources Office (PENRO), which exercises a number of devolved line functions, such as forest protection and titling, and the administration of small-scale mining. More influence is exercised less formally, however, through the provincial government's use of its superior analytical capacity and its ability to impose fiscal *quid pro quo* arrangements to guide and constrain municipal decision making.

Municipal governments are mandated to undertake water and soil resource utilization and conservation projects, implement community-based forestry projects, manage and control commercial forests with an area not exceeding 50 km^2, and establish tree parks, green belts and similar forest development projects. All these functions, however, are mandated *pursuant to national policies and subject to supervision and review of the DENR* (DILG, 1991, emphasis added). This illustrates that, by law, the national office still controls the environmental programmes of the local governments. Low-income upland municipalities typically do not have an Environment and Natural Resources Office (ENRO); the local government code provides for this only as an 'option.' In our study municipality, the *de facto* ENRO is a staff of the province. With two bosses, this person faces challenges in conducting his or her job effectively. On the other hand, all municipalities have a municipal agricultural officer (MAO). In many upland areas agriculture is the main driver of environmental change; hence the devolved MAO is, in practice, a key agent for environmental policy.

Further down the governance level, the village or *barangay* government has no legal support to conduct environmental programmes. Village government could ideally be an active proponent of some such initiatives, because it is closer to communities and households and thus better able to articulate local concerns. At present, village-level environmental programmes such as soil conservation, tree planting, monitoring buffer zones, and the like are initiated mainly by external agencies such as NGOs and externally funded projects in collaboration with national, provincial and municipal governments; village-level governments have no clear role except as cooperators. However, the long-term sustainability of externally designed and funded environmental initiatives may be heavily dependent on the extent to which they draw in, strengthen and confer 'ownership' to community and village institutions.

In an important recent innovation, the Philippines is now witnessing emergence of multi-level, multi-stakeholder institutions in which all relevant stakeholders – communities, local and national government – collaborate to address resource management issues. In theory at least, these have the potential to overcome many of the problems faced by individual local governments and line agencies, such as external costs, capacity constraints and overlapping mandates.

Decentralized Resource Management in the Manupali River Watershed

In this section we summarize a 9-year study (1994–2003) of resource management in Lantapan which illustrates the realities and challenges in upland management under decentralized regimes.

Early development

Lantapan's forestland was opened by commercial loggers granted a Timber License Agreement (TLA) by the national government. Agriculture followed the loggers, with migrants from other parts of the Philippines contributing to rapid area expansion. Large-scale deforestation took place from the 1950s to the 1980s. The uplands were indeed seen as a source of 'green gold' by lowlanders (Malanes, 2002), and the spread of intensive upland agriculture was largely driven by market opportunities (Coxhead *et al.*, 2001). Fallow periods were short, and in general, soil conservation was not practised, leading to significant land degradation. During 1974–1994, primary forest cover was reduced from half to less than one-third of municipal land area, being replaced mainly by maize and maize-based farming systems (Li Bin, 1994). Resource management decisions were the responsibility of national agencies (especially those now known as the Department of Agriculture and the Department of Environment and Natural Resources), although in practice, land use rights and practices were allocated largely through local and informal mechanisms. In-migrants (some relocated from Northern Luzon highland areas in government programmes) acquired land from indigenous people in exchange for small sums or in barter trade; ownership claims were established through officially invalid means such as land tax declarations (Paunlagui and Suminguit, 2001). In Lantapan, the transition from customary law to forest and agricultural land occurred in little more than two decades following Independence.

1990s to the present – transition to the late development period

Our data show that the deforestation rate in the Manupali River watershed has decreased in the past decade. The estimated rate of deforestation was about 0.6% annually during the period 1994–2001, less than half that of the 1.4% national estimate during the same period. Outmigration has begun to be observed, especially though not exclusively during the drought years 1997–1998 (Rola *et al.*, 1999). On the other hand, agricultural intensification continues to be driven by opportunities in domestic and international markets. Land use data also confirm the intensive cultivation mostly of annual crops not just in the lower slopes, but likewise in the upper watershed. Water quality monitoring since 1994 in several watersheds shows that measures of total suspended solids (TSS) are considerably higher in those areas where agricultural cultivation is more widespread, in spite of lower average slope, and that seasonal TSS peaks appear to coincide with months of intensive land preparation activity (Deutsch and Orprecio, Chapter 3).

On a wider watershed scale, other consequences of rapid and increasing soil erosion due to agricultural intensification can be seen in the deterioration of two water impoundment structures, the MANRIS diversion dam and the Pulangi IV hydropower facility, located a few kilometres below the junction of the Manupali and Pulangi rivers. Although forest management policies and strategies to reduce deforestation were adopted in the 1990s, and despite a ban on commercial logging, policies to promote sustainable upland management have yet to translate into better environmental health.

Institutional innovations under decentralization: the protected area management board

In the Philippines, local management of natural parks is stipulated in the National Integrated Protected Areas System (NIPAS) law. Mt Kitanglad Range Natural Park, a protected area, is situated in the Bukidnon provinces. Some management responsibility is accorded to the Protected Area Management Board (PAMB), whose members include representatives from line agencies, local government and the community. The Chair of the Board is the regional director of the DENR.[3] The PAMB regularly meets to provide oversight and guidance to field implementers.

Though a central government creation, the PAMB recognizes elements of customary law. A Council of Elders (COE) reinforces the natural park management by applying customary law in dealing with violators. The COE is stipulated in the Indigenous People's Rights Act (IPRA), under which, indigenous peoples are given authority to practice their customary rules in the management of natural resources. The Kitanglad Guard Volunteers (KGV), a people's organization, patrols and monitors illegal activities, with offenders undergoing a *sala*, or cleansing ritual. However, it has been observed that local tribal leadership through such institutions has emerged and begun to influence management of the protected area – sometimes extending beyond defined legal terms. As a result, officials see it as necessary to resolve and clarify the dividing line between legally instituted structures and traditional management structures (PAMB, undated).

Watershed Protection and Development Councils

Watershed Protection and Development Councils now exist for both the province and the study municipality. At the province, this is a multi-sectoral body composed of national and local agencies, academe, local government units and representatives from non-governmental organizations (NGOs). The 1995 presidential decree creating the provincial body stipulated that it was 'in order to fully protect and preserve the remaining forests in the Bukidnon watersheds and rehabilitate open areas within their headwaters' (see Sumbalan, 2001). Representatives of resource agencies in the province make up the Technical and Advisory Committee (TAC). This body has the potential to enable local actors and stakeholders to realign their efforts towards wider watershed level programmes.

At the municipal level, the Watershed Management Council was formed in part as a response to advocacy efforts of the SANREM CRSP research project. This is a multi-sectoral group comprised of representatives from agribusiness, NGOs, people's organizations, the municipal legislative council and provincial agencies. A Municipal Watershed Management Plan was recently approved for use as a guide to watershed management in the municipality. It is also an input into the provincial watershed management plan.

Institutional Challenges for Sustainable Upland Agriculture

In our study area, as in many other locations, the asynchronous development of markets and institutions has been associated with very rapid rates of forest conversion and land degradation. With growing market demand for resource-based products in the post-colonial era, the undermining or outright abrogation of customary laws with no effective replacement created a situation in which open access became widespread: the so-called 'period of disorder and destruction.' Rapid rates of deforestation and land conversion ensued. Subsequently, intensive agriculture became the leading culprit of upland destruction. During this period, national agencies increased their presence at the frontier, but the effective assertion of State control lagged far behind. Resource management policies began to be deconcentrated in the 1990s as part of a far broader set of decentralization measures. Moving decision making powers closer to the resources is generally seen as a good thing, but whether decentralization is a better alternative for watershed and natural resource management depends on whether institutional arrangements develop to overcome the inherent difficulties imposed by externalities and other constraints to effective local management.

The devolution of authority needs to be accompanied by institutional reforms for increased transparency and accountability and for building local capacity. The problem of jurisdictional boundaries and overlapping mandates must also be addressed. Multi-level, multi-stakeholder institutions such as the PAMB look like good first steps, but the specific design of such bodies must vary from place to place according to specific economic, cultural and institutional conditions, so it is too early to say that this model is a generalizable one. In increasingly highly organized market economies, the formal, corporatized private sector can and must be drawn into this process as an important and influential actor (Siamwalla, 2001). In the long run, protecting upland watershed and natural resources in decentralized resource management systems will depend on the sincerity of national governments in devolving functions, the competence of local governments to perform these functions, and the active participation of local community members and other local actors in the governance and management of their natural resource assets.

Notes

[1] Indonesia had imposed a similar logging ban in 1981, but the motivation was somewhat different, primarily aiming to protect and promote its domestic plywood industry.
[2] In the Philippines, a programme called the Industrial Forestry Management Agreement (IFMA) was implemented in 1991. This was a 25-year renewable contract between the Department of Environment and Natural Resources (DENR) and private business or legal commercial forest users, giving the latter rights to replant, and to harvest replanted trees (Vitug, 2000). After only limited success, the programme was suspended indefinitely in 1995. A similar programme in Indonesia (Timber Plantation Development) was also promoted, but its sustainability objective is falling short of expectations (Resosudarmo, 2002).
[3] This is a generic rule. Protected areas could be spread across more than one province. Mt Kitanglad range is situated in Bukidnon, and for this reason the provincial government actively participates in its management.

References

Asian Development Bank (1969) *Asian Agricultural Survey*. University of Tokyo Press, Tokyo.

Bromley, D.W. (1991) *Environment and Economy: Property Rights and Public Policy*. Basil Blackwell, Oxford, UK, and Cambridge, Massachusetts.

Burch, W.R. (1990) Foreword. In: Poffenberger, M. (ed.) *Keepers of the Forest. Land Management Alternatives in Southeast Asia*. Ateneo de Manila University Press, Quezon City, Philippines.

Colfer, C. and Resosudarmo, I. (eds) (2002) *Which Way Forward? People, Forests, and Policy Making in Indonesia*. Resources for the Future, CIFOR and Institute of Southeast Asian Studies, Singapore.

Coxhead, I. (2002) It takes a village to raise a Pigovian tax . . . or does it take more? Prospects for devolved watershed management in developing countries. Paper presented at a conference on Sustaining Food Security and Managing Natural Resources in Southeast Asia: Challenges for the 21st Century, Chiangmai, Thailand, 8–11 January 2002.

Coxhead, I. and Jayasuriya, S.K. (2003) *The Open Economy and the Environment: Development, Trade and Resources in Asia*. Edward Elgar, Cheltenham, UK, and Northampton, Massachussetts.

Coxhead, I., Rola, A.C. and Kim, K. (2001) How do national markets and price policies affect land use at the forest margin? Evidence from the Philippines. *Land Economics* 77, 250–267.

DILG (Department of Local Government) (1991) *Local Government Code of the Philippines*. Department of Local Government, Manila.

Dupar, M. and Badenoch, N. (2002) *Environment, Livelihoods, and Local Institutions: Decentralization in Mainland Southeast Asia*. World Resources Institute, Washington, DC.

FAO (Food and Agriculture Organization of the United Nations) (2001) *Global Forest Resources Assessment 2000*. FAO, Rome. Accessed November 2001 at: http://www.fao.org/forestry/fo/fra/main/pdf/main_report.zip

Fay, C. and Surait, M. (2002) Reforming the reformists in post-Suharto Indonesia. In: Colfer, C. and Resosudarmo, I. (eds) *Which Way Forward? People, Forests, and Policy Making in Indonesia*. Resources for the Future, CIFOR and Institute of Southeast Asian Studies, Singapore, pp. 126–143.

Gérard, F. and Ruf, F. (eds) (2001) *Agriculture in Crisis: People, Commodities and Natural Resources in Indonesia, 1996–2000*. CIRAD, Montpellier, France.

Gonzales, E.T. and Mendoza, M.L. (2002) Governance in Southeast Asia: issues and options. PIDS Research Paper No. 2002–06. Philippine Institute for Development Studies, Makati City, Manila.

Kummer, D. (1992) *Deforestation in the Postwar Philippines*. Ateneo de Manila Press, Manila.

Lewis, B. (2001) The new Indonesian equalization transfer. *Bulletin of Indonesian Economic Studies* 37, 325–343.

Li Bin (1994) The impact assessment of land use changes in the watershed area using remote sensing and GIS: a case study of the Manupali watershed, the Philippines. Unpublished master's thesis, Asian Institute of Technology, Bangkok.

Lynch, O.J. (1987) Philippine law and upland tenure. In: Fujisaka, S., Sajise, P. and del Castillo, R. (eds) *Man, Agriculture and the Tropical Forest: Change and Development in the Upland Philippines*. Winrock International, Bangkok, pp. 269–292.

Magno, F.A. (2003) Forest devolution and social capital state-civil society relations in the Philippines. In: Contreras, A. P. (ed.) *Creating Space for Local Forest Management in the Philippines*. CIFOR and La Salle Institute of Governance, Manila, Philippines.

Malanes, M. (2002) *Power from the Mountains: Indigenous Knowledge Systems and Practices in Ancestral Domain Management – The Experience of the Kankanaey-Bago People in Bakun, Benguet Province, Philippines*. ILO, Geneva, Switzerland.

Manasan, R. (2002) Devolution of Environmental and Natural Resources Management in the Philippines: Analytical and Policy Issues. *Philippine Journal of Development* 53, 33–54.

Manasan, R.G., Gonzalez, E.T. and Gaffud, R.B. (1999) Indicators of good governance: developing an index of governance quality at the LGU level. *Philippine Journal of Development* 48, 149–212.

PAMB (Protected Area Management Board) (undated). Mount Kitanglad Range Natural Park Management Plan. Department of Environment and Natural Resources, Malaybalay, Bukidnon, Philippines.

Paunlagui, M.M. and Suminguit, V. (2001) Demographic development in Lantapan. In: Coxhead, I. and Buenavista, G. (eds) *Seeking Sustainability: Challenges of Agricultural Development and Environmental Management in a Philippine Watershed*. PCARRD, Los Baños, Philippines, pp.138–160.

Poffenberger, M. (1990) The evolution of forest management systems in Southeast Asia. In: Poffenberger, M. (ed.) *Keepers of the Forest. Land Management Alternatives in Southeast Asia*. Ateneo de Manila University Press, Quezon City, Philippines.

Poffenberger, M. (ed.) (2000) *Communities and Forest Management in Southeast Asia: A Regional Profile of the Working Group on Community Involvement in Forest Management*. IUCN, Geneva, Switzerland.

Resosudarmo, I.A.P. (2002) Timber management and related policies. In: Colfer, P.

and Resosudarmo, P. (eds) *Which Way Forward? People, Forests, and Policy Making in Indonesia*. Resources for the Future, CIFOR and Institute of Southeast Asian Studies, Singapore.

Rola, A.C. and Coxhead, I. (2002) Does non-farm job growth encourage or retard soil conservation in the uplands? *Philippine Journal of Development*, 53, 55–83.

Rola, A.C. and Coxhead, I. (2005) Economic development and environmental policy in the uplands of Southeast Asia: challenges for policy and institutional development. In: Colman, D. and Vink, N. (eds) *Reshaping Agriculture's Contribution to Society: Proceedings of the Twenty-Fifth International Conference of Agricultural Economists*. Blackwell, Malden, Massachussetts, and Oxford, UK, pp. 243–256.

Rola, A., Tabien, C.C.O. and Bagares, I.B. (1999) Coping with El Nino, 1998: an investigation in the upland community of Lantapan, Bukidnon. ISPPS Working Paper 99–03. University of the Philippines at Los Baños, College, Laguna, Philippines.

Rola, A.C., Elazegui, D.D., Foronda, C.A. and Chupungco, A.R. (2003) The hidden costs of bananas: imperatives for regulatory action by local governments. CPAf Policy Brief. No. 03–01. College of Public Affairs, University of the Philippines at Los Baños, College, Laguna, Philippines.

Rola, A., Deutsch, W., Orprecio, J. and Sumbalan, A. (2004) Water resources management in a Bukidnon subwatershed: what can community generated data offer? In: Rola, A., Francisco, H. and Liguton, J. (eds) *Winning the Water Wars: Watersheds, Water Policies and Water Institutions*. Philippine Institute for Development Studies, Makati City, Philippines and Philippine Council for Agriculture, Forestry, and Natural Resources Research and Development, Los Baños, Laguna, Philippines.

Siamwalla, A. (2001) *The Evolving Roles of State, Private and Local Actors in Rural Asia*. Oxford University Press for the Asian Development Bank, New York.

Sumbalan, A.T. (2001) The Bukidnon experience on natural resource management decentralization. Paper presented at the SANREM conference, 'Sustaining Upland Development in Southeast Asia: Issues, Tools and Institutions for Local Natural Resource Management' 28–30 May 2001, ACCEED Conference Center, Makati City, Philippines.

Tachibana, T., Nguyen, T.M. and Otsuka, K. (2001) Management of state land and privatization in Vietnam. In: Otsuka, K. and Place, F. (eds) *Land Tenure and Natural Resource Management: A Comparative Study of Agrarian Communities in Asia and Africa*. Johns Hopkins University Press for the International Food Policy Research Institute, Washington, DC, pp. 234–272.

TDRI (Thailand Development Research Institute) (1994) *Assessment of Sustainable Highland Agricultural Systems*. Thailand Development Research Institute, Bangkok.

Vitug, M.D. (2000). Forest Policy and National Politics. In: Utting, P. (ed.) *Forest Policy and Politics in the Philippines: The Dynamics of Participatory Conservation*. Ateneo de Manila University Press, Manila, pp. 11–39.

World Bank (1995) *Mainstreaming the Environment: A Summary*. The World Bank, Washington, DC.

World Bank (2000a) *Indonesia: Public Spending in a Time of Change*. Accessed at http://lnweb18.worldbank.org/eap/eap.nsf August 2003. The World Bank, Washington, DC.

World Bank (2000b) *Thailand: Public Finance in Transition*. Accessed at http://www.worldbank.or.th/economic/index.html August 2003. The World Bank, Washington, DC

3

Water Quality Changes in the Manupali River Watershed: Evidence from a Community-based Water Monitoring Project*

W. G. DEUTSCH[1] AND J. L. ORPRECIO[2]

[1]International Center for Aquaculture and Aquatic Environments, Department of Fisheries and Allied Aquacultures, Auburn University, Auburn, AL 36849, USA, e-mail: deutswg@auburn.edu; [2]550 Mehan Street, Parkshomes Subdivision, Muntinlupa City, Philippines, e-mail: jorprecio@amore.org.ph

Introduction

Over the past few decades, there has been a growing interest worldwide in getting personally involved with environmental issues. Many community-level groups have sought ways to gather environmental data by themselves, without relying on 'experts' who are often unavailable, too expensive or unwilling to get involved with a local issue. Water quality and quantity issues are becoming increasingly important and are central to both ecosystem health and human development. Because of recent advances in simple technologies for testing water, local groups of non-specialists can now collect reliable information that is useful for watershed management, and is often the only empirical data available for decision makers and resource managers.

Human-altered land ultimately affects the quantity and quality of water. Rainfall percolates through soil or flows on the land's surface, and carries both suspended and dissolved substances to groundwater and streams. How soil is covered and protected affects the amount and pattern of run-off, erosion and sedimentation. Likewise, the types of household and agricultural chemicals used, and how human and livestock wastes are disposed of, will eventually impact water and water users. Land-based activities are manifested in water and determine the aesthetic and productive characteristics of watersheds.

*We thank Vincent Molina, President of the *Tigbantay Wahig*, and all members of CBWM groups in the Philippines for devoting thousands of hours of volunteer time for water monitoring and community service related to this project. Janeth Labis, Wendell Talampas, Eric Reutebuch and Sergio Ruiz-Córdova assisted with data management and analyses.

Many communities in the developing world share similar accounts of how water quality and quantity has degraded in living memory. Community elders will recount how stream water was drinkable and stream flow was steady in the past, whereas it is presently polluted and undergoes more frequent flooding and drought cycles. These locally perceived changes in watershed function provide an impetus for community action and, when coupled with proper training in use of simple test kits, can lead to important data for environmental restoration and protection.

For the last 14 years, several research projects have worked in varying degrees of integration to understand and quantify ecological, social and economic conditions and trends in the Manupali River watershed and its various subwatersheds (see Plate 4). This chapter reports findings from a community-based water monitoring (CBWM) project that was conducted between 1993 and 2003 in the watershed. At the outset, it is important to underscore that the activity was not strictly limited to water quality monitoring and data collection. We aimed not only to collect important and credible information about changes in water quality, but also to empower various community groups to engage in environmental assessment and influence trends in water quality and quantity. Below we provide an overview of how the CBWM effort was implemented in the watershed, report evidence on water quality and discuss the essential features of a viable CBWM group. Throughout, we highlight the importance of four cornerstones of research that were established at the start of the project: (i) participation, with substantial inputs from local community members in the design and implementation of the research; (ii) interdisciplinary activity, with research partners representing distinct and complementary specialities; (iii) intersectoral linkages, with two or more components of society (agriculture, education, business, government, religion, etc.) involved in the project; and (iv) landscape scale, with research activities occurring in at least two agroecological zones within the watershed (Foglia, 1995).

Whereas some water quality monitoring projects invest in highly quantitative and expensive equipment for environmental assessment, our project invested in enhancing local capabilities for monitoring water with relatively simple tools. The CBWM project had much participation from the local community, with a balance of men and women attending the workshops and becoming water monitors. Research partners included aquatic ecologists, biologists, educators and development specialists. Local groups involved in the project included schools, provincial, municipal and village-level (*barangay*) government officials, and farmer and tribal groups. Monitoring activities have occurred on a landscape scale and have encompassed virtually all agroecological zones of the Manupali River watershed, ranging from areas of primary forest to those consisting mainly of lowland rice farms.

Methods and Approach

Prior to implementing the water quality monitoring project, an extensive community assessment, called the Participatory Landscape/Lifescape Appraisal (PLLA), was conducted to identify research priorities in the watershed (Bellows *et al.*, 1995). It was clear during the PLLA that water-related issues were at the forefront of community concerns. These concerns included drinking water quality, flooding and drought cycles, and a general sense of watershed degradation over previous decades. During 1993 and early 1994 the CBWM project conducted demonstrations of water testing and carried out some preliminary assessments of streams with local farmers, educators and non-governmental organization (NGO) representatives.

Many of the techniques and training materials for the project came from the Alabama Water Watch programme, funded by the US Environmental Protection Agency (EPA) and the Alabama Department of Environmental Management to promote the development of CBWM in the south-eastern USA (Deutsch *et al.*, 1998). These monitoring protocols and training resources were vital for implementing the project in the Philippines, but needed to be significantly adapted to the local situation. Different styles and lengths of workshops, translations of materials into local dialects and monitoring new water parameters were all part of this adaptive process.

An initial water quality training workshop took place in Lantapan in July 1994. Many of the first participants were farmers who had been previously involved with sustainable agriculture projects. Other participants were associated with previous projects of a key partner in the research, Heifer International, Philippines (HI/P). Some trainees were members of the local *Tala-andig* tribal community, the indigenous group that claims ancestral land rights to much of the Manupali River watershed.

The training workshop lasted three days and included extensive hands-on activities in addition to classroom training. The basic approach was to introduce concepts of watersheds and environmental management, then provide instruction on the use of water monitoring equipment (see Plate 5 for a photograph of *Tigbantay Wahig* members taking water samples). Water quality test kits used in the project were portable and intended for field use by non-specialists. The water chemistry test kit had been custom-made for Alabama Water Watch, and included supplies for measuring water temperature, pH, alkalinity, hardness, dissolved oxygen and turbidity. Test kit results previously had been compared with those of Standard Methods analyses and were found to be within acceptable limits of bias and precision. As a result, the water chemistry protocols had previously received EPA approval for use by CBWM groups in Alabama (Deutsch and Busby, 1994, 1999).

A second part of the first workshop was to conduct training in the measurement of total suspended solids (TSS). This water quality variable was of particular interest to both researchers and farmers because of its obvious link to soil erosion and sedimentation. Erosion of farmland on steep slopes and siltation of streams, irrigation canals and a local hydro-power reservoir were commonly recognized as environmental problems that had direct impacts on people's lives. A research partner at nearby Central Mindanao University oversaw the processing of filters needed to collect the TSS samples, and submitted the data (pre- and post-weights of filters) to HI/P for storage in a database and electronic transmission to Auburn University (AU).

By the end of the 3-day session we were able to certify about 15 community members as water monitors. They then began a monthly monitoring programme on several sites of four main rivers in the municipality, namely the Tugasan, Maagnao, Alanib and Kulasihan rivers (Deutsch and Orprecio, 2000). The HI/P programme provided staff for technical support of the new group which named itself the *Tigbantay Wahig* (literally meaning 'water watchers' in the local *Binukid* dialect). Later, HI/P expanded its field office in Lantapan and built a training centre to facilitate the numerous, additional workshops that were conducted over the subsequent decade to train other water monitoring groups.

The consistent collection of data by the *Tigbantay Wahig* (TW) resulted in the first, systematic study of water quality and quantity in Lantapan (Deutsch *et al.*, 2001a). Project partners regularly met with the TW to discuss the meaning of the data and to refine the monitoring plan. Over the following 1–3 years, training was provided for measurement of additional water variables. These included: (i) stream bio-monitoring using aquatic invertebrates; (ii) measurement of *Escherichia coli* and other coliform bacteria in water; and (iii) estimates of stream current, discharge and sediment yield. Monitoring of water chemistry, TSS, bacterial concentrations and invertebrates focused on analyses of water quality. The stream current and discharge measurements were related to water quantity, and provided indicators of the amount and variability of stream flow.

As the technical skills of the TW grew, so did their organization and leadership. HI/P provided companion workshops for developing the group, helping them plan and professionalize. Examples included leadership training, accounting and bookkeeping, and establishing a mission and goals. Because the water monitoring activity was perceived as relevant to community service and family health as well as being enjoyable, the TW grew in numbers and influence. The group incorporated as a registered people's organization in 1995, and subsequently established by-laws and held annual elections of officers (Deutsch *et al.*, 2001b).

Results of a Comprehensive Watershed Assessment

Between 1993 and 2004 the TW group analysed thousands of water samples for various physical, chemical and biological characteristics. The resulting datasets elucidate watershed trends on a landscape scale. Significant findings may be divided into four components: (i) physico-chemical tests; (ii) TSS; (iii) bacteriological (*E. coli*) monitoring; and (iv) stream discharge and sediment yield.

Physico-chemical tests

Six physico-chemical variables were measured during a sampling session using a portable test kit (Table 3.1). All tests were conducted streamside, and required about 30 min to complete. Each of the six variables was measured monthly in each of the four main rivers draining the subwatersheds (see Plate 4) from 1994 to 2003. This resulted in more than 2000 total measurements. An analysis of variance (ANOVA) was used to determine statistical differences of each variable among the four rivers. A Tukey's studentized test (multiple comparison procedure) was used to indicate which streams differed significantly (Cody and Smith, 1997).

Water temperatures throughout the Manupali River watershed ranged from about 15°C to 35°C (Fig. 3.1). The average temperature of the Tugasan and Maagnao rivers was about 20°C, with similar ranges. The average temperatures of the Alanib and Kulasihan rivers were significantly higher than in the western two subwatersheds by about 3°C and 5°C, respectively (Table 3.1). The progressive warming of the rivers across the landscape was probably influenced by two factors. The elevation of the four, primary sampling sites decreased by about 1000 m from west to east, with increasing air temperatures. Also, the two eastern subwatersheds were considerably less forested and, therefore, had more direct solar radiation to the stream channels.

The average dissolved oxygen (DO) of the streams ranged 7.7–8.0 mg/l and, in most cases, was supportive of fish and other aquatic life. A DO of at least 5.0 mg/l is usually required to maintain a healthy fishery. There was a slight reduction in mean oxygen levels in the eastern watersheds (Fig. 3.1), which may have been the result of slightly higher water temperatures. The

Table 3.1. Measurements for six physico-chemical variables of the Tugasan, Maagnao, Alanib and Kulasihan rivers, Lantapan, Bukidnon, Philippines, 1994–2002.

Parameter	Statistic	Tugasan (T-1)	Maagnao (M-2)	Alanib (A-3)	Kulasihan (K-3)
Water temperature (°C)	average	20.4[a]	20.0[a]	22.9[b]	25.3[c]
	range	(15–28)	(15–30)	(16–35)	(19–31)
	N	77	79	80	76
Dissolved oxygen (mg/l)	average	8.0[ab]	8.1[a]	7.8[ab]	7.7[b]
	range	(4.6–9.2)	(6.0–9.9)	(6.6–9.7)	(3.8–9.2)
	N	70	75	74	64
Total alkalinity (mg/l)	average	41.2[a]	59.5[b]	56.5[b]	32.2[c]
	range	(20–70)	(25–115)	(20–100)	(15–100)
	N	78	81	80	75
Hardness (mg/l)	average	36.6[a]	55.7[b]	49.8[c]	25.4[d]
	range	(5–70)	(30–100)	(5–80)	(5–60)
	N	77	81	78	74
pH (SU)	average	7.6[a]	7.8[b]	7.7[ab]	7.1[c]
	range	(7.0–8.5)	(7.0–8.50)	(7.0–8.5)	(5.0–8.5)
	N	77	77	78	75
Turbidity (JTU)	average	5.5[a]	6.3[a]	10.0[a]	20.8[b]
	range	(0–20)	(0–40)	(0–40)	(5–70)
	N	61	66	70	66

Note: Averages in a row with different letters are significantly different at the $\alpha = 0.05$ level.

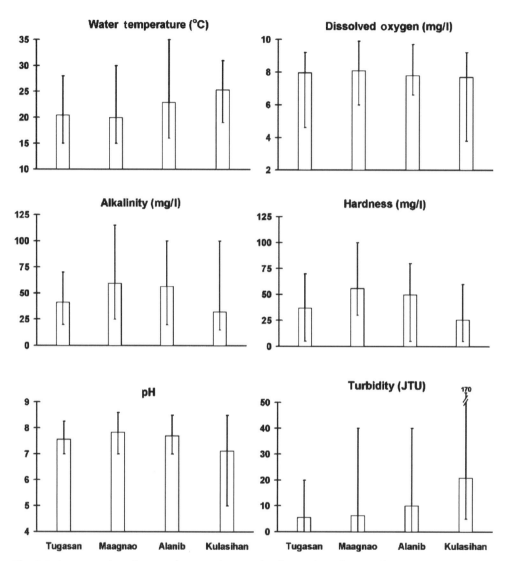

Fig. 3.1. Average values (bars) and ranges (lines within bars) of six physico-chemical variables of the Tugasan, Maagnao, Alanib and Kulasihan rivers, Lantapan, Bukidnon, Philippines, 1994–2002.

solubility of oxygen in water is inversely related to temperature. Also, there are more humans and livestock in the eastern subwatersheds and they probably contributed more organic pollution (sewage, household waste) that would consume oxygen in streams. Although the 10-year averages were similar, the DO of the Maagnao River was significantly higher than that of the Kulasihan River (Table 3.1).

The pH, total alkalinity and total hardness of water are interrelated variables which had a similar pattern in the watershed (Fig. 3.1). Overall, water quality in the four rivers may be characterized as mildly alkaline, with moderate buffering capacity and hardness. Average values of the three variables were usually significantly higher in the middle two subwatersheds (Table 3.1). This observation is probably related to the geology and soil formation of the region (Poudel and West, 1999), derived from a mixture of rocks of volcanic origin (mainly basalt) and elevated seabeds (mainly limestone). Results suggest that water quality in the Maagnao and Alanib subwatersheds is proportionately more affected by limestone deposits. Limestone substrates (calcium carbonate) contribute both calcium (raising total hardness) and carbonates (raising alkalinity) to a stream and the latter is associated with an elevated and more stable pH.

Turbidity readings, measured in Jackson Turbidity Units (JTUs), sharply increased from west to east among the four subwatersheds. The turbidity of the Kulasihan River was significantly higher than in the other three rivers (Table 3.1), and averaged about five times higher than in the Tugasan River (Fig. 3.1). This indicator of water clarity is usually directly related to soil erosion, especially in mountain streams such as in Lantapan. Turbidity readings are also directly related to measurements of TSS, and the TW data substantiated this relationship.

Total suspended solids

The multi-year pattern of monthly TSS measurements collected during base flow and total monthly rainfall in the Maagnao and Kulasihan rivers may be seen in Fig. 3.2. The average annual rainfall collected at two weather stations in the Manupali River watershed was similar and ranged from 2245 mm in the Kulasihan River subwatershed to 2366 mm in the Alanib River subwatershed (Table 3.2). The amount of TSS was much higher in the Kulasihan River than in other streams, and this supported the general observation of the

Table 3.2. Rainfall at the Alanib and Kulasihan weather stations, Lantapan, Bukidnon, Philippines, 1994–2002.

Year	Total annual rainfall (mm)	
	Alanib	Kulasihan
1994	2076	2113
1995	3007	2991
1996	2294	2734
1997	1714	1569
1998	2010	1337
1999	2716	2566
2000	2856	2535
2001	2258	2113
Average	2366	2245

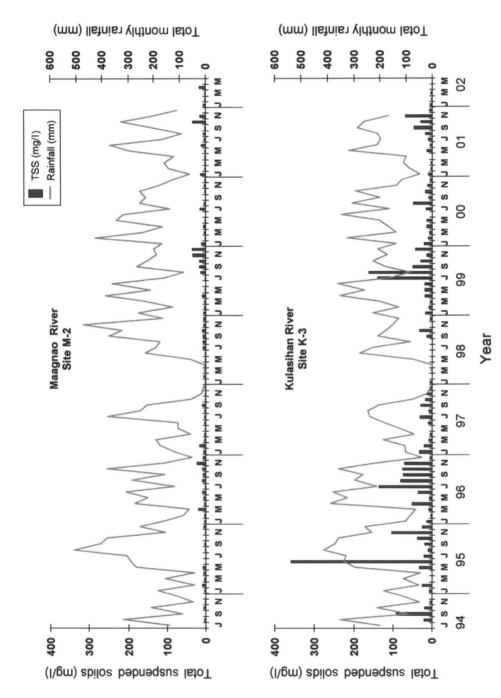

Fig. 3.2. Monthly average total suspended solids and total monthly rainfall for the Maagnao and Kulasihan rivers, Lantapan, Bukidnon, Philippines, 1994–2002.

community that the river was frequently 'muddy'. Because the TSS readings were made only once per month in base flow conditions, correlations with rainfall were not strong. However, from 1995 to 2001, there was a significant ($P = 0.03$) correlation between the total annual rainfall and average annual TSS in the Kulasihan River and a general tendency for TSS to be less in all streams during periods of less rainfall. During the El Niño drought of late 1997 and early 1998, both rainfall and TSS in all streams neared zero (Fig. 3.2).

Figure 3.3 depicts two summary graphs of TSS from the four subwatersheds. The first was based on about 160 samples collected during the first few months of the project. The second graph is based on nearly 1350 samples collected from the same streams over a 9-year period. The two graphs tell essentially the same story. The amount of TSS, as an indication of soil in run-off water to streams, progressively increased moving west to east across the four subwatersheds. The two western subwatersheds had considerably more

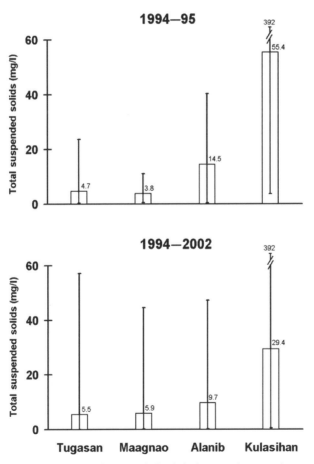

Fig. 3.3. Average concentration of total suspended solids (bars) and ranges (lines within bars) in the Tugasan, Maagnao, Alanib and Kulasihan rivers, Lantapan, Bukidnon, Philippines, 1994–1995 and 1994–2002.

forest cover and lower human population density than the two eastern subwatersheds (Bin, 1994; Paunlagui and Suminguit, 2001). Results reveal that soil erosion from areas including agricultural land, clear cuts, construction sites and stream banks was greater in the more developed portions of the Manupali River watershed (Deutsch *et al.*, 2001a).

Perhaps equally importantly, the TSS results demonstrated that a community group was motivated to systematically monitor and obtain such valuable information, conducting most field sampling without outside intervention. The environmental gradient of TSS, generally described by the TW in the first few months of sampling, also suggested that broad-based environmental assessments may be done by community groups in a cost-effective way over a relatively short period.

More subtle changes of TSS within the subwatersheds were quantified by continued monitoring over several years. Although there were no significant differences in the average annual TSS of the Tugasan, Maagnao and Alanib rivers from 1994 to 2003, concentrations of TSS in the Kulasihan River significantly ($P = 0.05$) declined (Fig. 3.4).

The apparent decrease of TSS in the Kulasihan River may have resulted from several factors. One contributing factor was the series of droughts that occurred in 1997/1998 and 2000/2001. These dramatically reduced rainfall, erosion and stream flow in the Kulasihan. A second factor may be local ordinances enacted in 2000 by the Municipality of Lantapan that aimed to create a larger buffer zone along the banks of the river. Such buffers filter and settle sediments before they enter streams. A third factor was the Kulasihan River Restoration Project, which was initiated by the municipality and the TW in 2002 to plant hundreds of giant bamboo to reduce bank erosion and improve the stream-side buffer. A final mitigating factor may have been the natural and human-induced re-vegetation of cleared forest and agricultural land that could have decreased erosion rates in the Kulasihan River subwatershed.

The base flow TSS measurements established a relative comparison of conditions and trends among the four subwatersheds. To expand the understanding of erosion and stream sedimentation, TSS was monitored during several rainfall events. During these periods, TSS concentrations were 7 to 36 times higher than in base flow conditions (Table 3.3).

Bacteriological (*E. coli*) monitoring

The primary bacteriological data collected during the project have been presented in Deutsch *et al.* (2000; 2001a). Four extensive surveys in 1995/1996 (about 150 samples per survey) revealed a pattern of degradation in surface water, from west to east, that was similar to that observed for TSS (Figs 3.5 and 3.6). The municipal drinking water is primarily distributed by gravity through pipes from mountain springs, and monitoring of water at community taps revealed a general pattern of increased bacterial contamination related to distance from spring source. This was probably because of deterioration of the

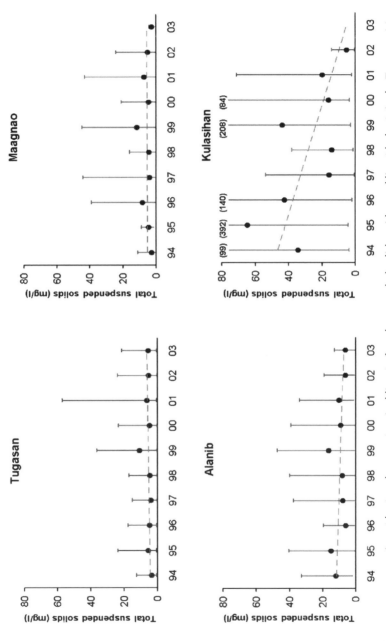

Fig. 3.4. Average concentrations (dots) and ranges (vertical lines) of total suspended solids, with trend line (dashes), in the Tugasan, Maagnao, Alanib and Kulasihan rivers, Lantapan, Bukidnon, Philippines, 1994–2003.

Table 3.3. Total suspended solids (mg/l) in the Tugasan, Maagnao, Alanib and Kulasihan rivers, Lantapan, Bukidnon, Philippines, 1994–2002.

Stream	Regular sampling				Rain event sampling			
	N	Min	Max	Average	N	Min	Max	Average
Tugasan (T-1)	345	0.1	57	5	369	0.9	1652	106
Maagnao (M-2)	355	0.2	45	6	402	0.3	1088	85
Alanib (A-3)	373	0.2	47	9	435	0.9	17,540	323
Kulasihan (K-3)	297	0.3	392	28	326	0.6	3400	208

Note: Average concentration of total suspended solids, measured at base flow and during rainfall events.

piping system and increased likelihood of bacteria entering the system through breaks in pipes, then moving downstream. Partly as a result of these surveys, the Municipality of Lantapan secured national government grants to refurbish the piped drinking water system of the municipality.

Stream discharge and sediment yield

Dramatic differences in stream discharge patterns were found between the Maagnao and Kulasihan rivers, based on five years of monthly monitoring (Fig. 3.7). The Maagnao River had relatively stable flow, even during severe

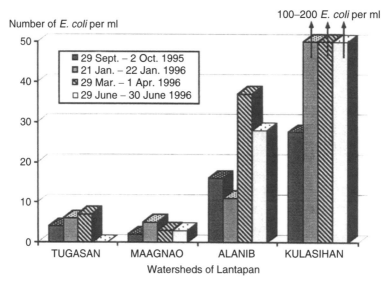

Fig. 3.5. Average concentration of *Escherichia coli* bacteria in four surveys of the Tugasan, Maagnao, Alanib and Kulasihan rivers, Lantapan, Bukidnon, Philippines, 1995–1996 (Deutsch *et al.*, 2001a).

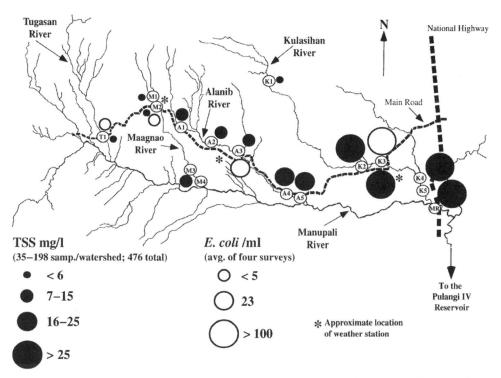

Fig. 3.6. The Manupali River watershed with estimates of total suspended solids and *Escherichia coli* bacteria concentrations, Lantapan, Bukidnon, Philippines, 1994–1997. (*Source*: Deutsch *et al.*, 2001a.)

droughts, ranging about 1–3 m³/s. In contrast, the Kulasihan River was very unstable in its discharge, ranging 0–10 m³/s. The discharge of the Kulasihan River was largely influenced by rainfall events. The TW data revealed that the Kulasihan River, though rural, responded to rainfall much like an urban, 'flashy' stream. This was probably because the subwatershed is mostly cleared of forests and has relatively little infiltration of rainfall to ground water.

Stream discharge data were used with concurrently measured TSS data to make sediment yield estimates. This variable is the estimated dry weight of suspended solids (primarily eroded soils) flowing past a given stream site per unit time. As expected, the Kulasihan River had more sediment yield than more forested subwatersheds to the west, and periods of high sediment yield were generally associated with high stream discharge. Sediment yield was 1400–3400 kg/h on four occasions between 1997 and 1999 on the Kulasihan River, but never exceeded 400 kg/h on the Maagnao River during this same period (Fig. 3.8).

Stream discharge estimates were also converted to specific discharge (m³ of water/s/ha of watershed), based on subwatershed area upstream of the sampling site. Data indicated that the specific discharge was considerably higher in the relatively smaller Maagnao River subwatershed (about 5000 ha)

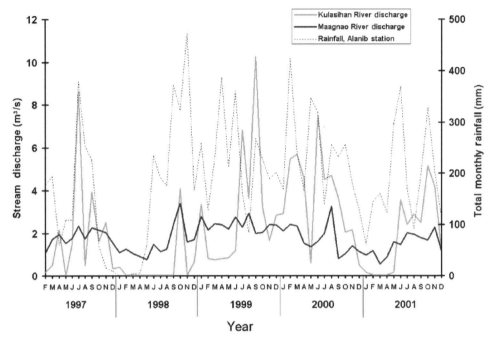

Fig. 3.7. Stream discharge of the Kulasihan and Maagnao rivers, with total monthly rainfall, Lantapan, Bukidnon, Philippines, 1997–2001.

than in the Kulasihan River subwatershed (about 10,000 ha). The relatively low specific discharge of the Kulasihan River can potentially be traced to three factors: (i) upstream water withdrawals for irrigation or municipal drinking water; (ii) unaccounted-for, subterranean flows of water; and (iii) the relatively infrequent (once per month) sampling of stream discharge by the TW. The latter explanation is probably the main reason why specific discharge calculations for the Kulasihan River were so low. With a monthly sample, there would be a much higher probability of measuring a representative discharge in the more stable Maagnao River than there would be in the highly variable flow of the Kulasihan River.

A complete understanding of stream discharge patterns requires sophisticated and expensive equipment to obtain instantaneous stream flow or depth measurements. An attempt to describe the flow of the rivers of Lantapan more quantitatively was made using existing data of soils, land slope and cover, rainfall patterns and the WEPP Model (see Chapter 7). The model used is accepted as state-of-the art for streams, but requires large amounts of various data to be run with minimal assumptions. In the case of Lantapan streams, instantaneous stream flow measurements were unavailable, and the model was, therefore, run making several assumptions. Results differed greatly from the estimated hydrographs of the TW monitoring group (Fig. 3.7). In particular, the maximum discharge of the Maagnao River was predicted to be much greater than the observed values of the TW, based on once per month sampling. It is

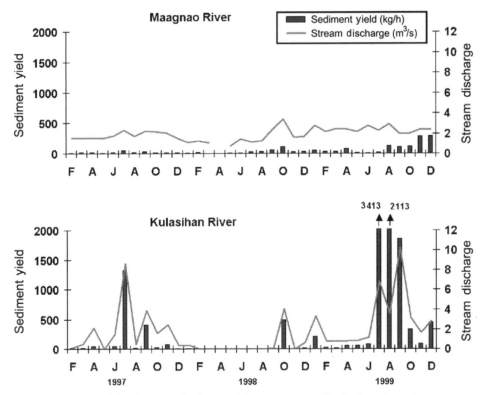

Fig. 3.8. Sediment yield and stream discharge of the Maagnao and Kulasihan rivers, Lantapan, Bukidnon, Philippines, 1997–1999. (*Source*: Deutsch *et al.*, 2005.)

still undetermined which method is the more accurate, though both clearly have advantages and limitations.

Summary of key findings of the watershed assessment

As the data set on water quality measurements expanded, the TW and other community members came to be able to consider conditions and trends in water quality in Lantapan and to begin to understand complex watershed processes. The water monitors now had locally generated information to document several factors which affected water quality and quantity. These included:

1. Seasonal fluctuations in water quality and quantity – some variables were influenced by seasonal temperature changes and TSS concentrations were considerably higher in the rainy season than in the dry season.

2. Trends and inter-year variations in water quality and quantity – data from a given site over several years facilitated trend analyses to answer the basic question, 'Is the stream getting better or worse?'

3. Upstream/downstream differences in water quality and quantity – monitoring a longitudinal gradient identified 'hotspots,' or critical areas of a stream where water quality degraded because of local pollution or changes in water flow.
4. Inter-watershed differences in water quality and quantity – there were distinct differences among the four, contiguous watersheds that related, in part, to elevation, geological differences and land use.
5. The role of climatic events in influencing water quality and quantity – several years of stream discharge measurements documented two El Niño events which had dramatically different effects on subwatersheds relative to land cover.
6. The role of human land use in determining water quality and quantity – the way in which the land in various subwatersheds was used had a clear effect on water quality and quantity. Good water quality was generally correlated with low human population, high percentage of forest cover and relatively little agriculture.

It is important to note that many of these factors work in combination and lead to the possibility that erosion 'hotspots' may exist in both space and time. In Chapter 8, Paningbatan undertakes a more formal analysis of what might constitute erosion 'hotspots' in the Manupali River watershed.

Discussion of Process Issues

The comprehensiveness and quality of the information collected by non-specialists attracted the attention of local residents, development specialists and policymakers. The reputation of the TW enabled them to initiate various action strategies to disseminate their watershed information and effect positive change. Human response to the water quality data was a major consequence of the data collection effort. Four basic responses and strategies to address water quality issues were implemented in the watershed. These are briefly highlighted below.

Environmental education

The TW began to educate teachers and students about the conditions of streams in Lantapan. Some teachers became certified water monitors and brought their students to workshops and field sessions. An environmental pen pal programme was initiated, wherein classrooms in the Philippines and the USA were partnered for student correspondence about water and the environment. These and other environmental education activities were endorsed by the Philippine Department of Education, Culture and Sports, for expansion and replication.

Stream protection and restoration

After the TW data documented the environmental degradation of subwatersheds in Lantapan, the Municipal Council established a Watershed Council to protect clean water and attempt to correct known problems. A large sign was posted at the bridge crossing of the Tugasan River, indicating that it was one of the cleanest rivers in Region X of the Philippines. The TW and HI/P were appointed by the council to oversee the planting of hundreds of giant bamboo seedlings on the eroded banks of the Kulasihan River.

Advocacy and policy

The TW have consistently provided community feedback of their data and expressed their positions regarding better water policy. Numerous municipal and *barangay* meetings have been attended by the TW to promote their views of clean water. Representatives of the group addressed the Philippine Congress and the TW has become a model of community participation for addressing watershed issues. The approach and results of the CBWM work plan has been developed into policy briefs for presentation to policymakers at the federal level. These briefs may be instrumental in drafting the citizen participation portions of the Clean Water Act of the Philippines (Rola, 2004).

Dissemination of information and the CBWM approach

The TW group has grown in numbers and has inspired the formation of new groups. Through networks of HI/P and other NGOs, numerous study tours have brought interested people to Lantapan to see various demonstrations of TW sampling techniques and group operation. Active water monitoring groups have now become established in three provinces of Mindanao (Bukidnon, Sarangani and Baungon) and in several municipalities of the island province of Bohol.

Conclusions and Directions for Future Work

Overall, the investment in CBWM has resulted in community group formation, customized, water monitoring techniques, training materials and workshops, credible, long-term data sets that describe watershed trends and a variety of successful action strategies. Several factors have contributed to the development of CBWM in the Manupali River watershed. These same factors are at play to varying degrees in other locations in the Philippines and elsewhere. Among the most important factors is the continued decentralization of authority in natural resources management. This is part of a worldwide trend wherein responsibilities and funding for protecting the environment are being transferred from national to regional and local levels in many countries. Accompanying this process is the increased receptivity to the concept of community-based environmental assessment and management.

Another positive contributor to the success of CBWM is the growing awareness that many people have of the environmental movement. Whether by formal educational programmes, popular media or personal experience of environmental problems, most people realize the importance of protecting forests, water and other resources. Water issues have been primary in recent international and regional conferences and in meetings of world leaders. The lack of good quality water in adequate quantity limits development and quality of life for millions. Governments are realizing that regulations alone are inadequate to solve water problems, and local efforts to address water issues are increasingly recognized as a vital part of the solution. An important consequence of the CBWM effort in the Manupali River watershed is that local groups are in a position to contribute to ongoing debates regarding protection of water resources.

After a decade of CBWM project development, there are several questions to face when forming a vision for the future. Some of the primary questions are: (i) What are the necessary organizational and technical characteristics of a water monitoring group to ensure it remains viable over time? (ii) How does a group become locally or regionally institutionalized? (iii) How can CBWM groups better contribute to natural resource management and improve a community's quality of life? and (iv) What is the best type of regional, national or international CBWM programme to foster the creation, development and networking of groups?

The Manupali experience suggests that in order to become and remain effective, a local group must have the technical skills to maintain a reputation for collecting quality data. This not only includes the ability of existing monitors to follow protocols and maintain testing equipment, but must also include an adequate number of certified, local trainers to conduct future workshops. Training-of-trainer workshops have been conducted over the last few years in the Philippines, and there are now about 15 certified CBWM trainers countrywide. These trainers receive support from the NGO partner, HI/P, and training workshops are primarily conducted with HI/P staff. This arrangement provides for systematic encouragement and refinement of training skills under the supervision of professionals, and exposes new monitors to veteran monitor-trainers from their community. Workshops and informal interactions among trainers and trainees may thus be conducted in local dialects for clearer communication of principles and techniques.

Another component that contributes to the effectiveness and longevity of a CBWM group is the establishment and maintenance of links to government and the business community. Endorsement of group activities and recognition of the value of the data by governmental agencies greatly facilitates the establishment of the activity and the use of the information. This partnership will affect the type, amount and location of water data collected. The information will then be of maximum benefit to local policymakers and future monitoring can adjust to goals and standards of environmental protection. In the case of the Manupali River watershed, the municipal government of Lantapan has demonstrated its willingness to use the TW data and allow TW members to participate in Natural Resource and Watershed Councils. The municipal government of Maitum, Sarangani, has provided training opportunities for members of its staff and has financially supported the monitoring activities of the CBWM group there, called the Munong El. At the federal level, representatives of the Environmental Management Bureau of the Department of Environment and Natural Resources have expressed interest in the CBWM model of Lantapan for possible replication throughout the country.

Finally, links to business and industry are valuable for a CBWM group because agricultural production practices have potentially negative environmental impacts and a CBWM group provides a method for facilitating feedback and exchange between a community and business interests. In the case of Lantapan, a local banana plantation, Mount Kitanglad Agro-Ventures, Inc. (MKAVI), has interacted with the TW regarding water withdrawals from the Maagnao River and pollution from plantation effluents. The company has since requested stream discharge data from the TW as an input to its plantation management plan, and has agreed to finance testing equipment for the group.

The above examples begin to describe a sustainable group that remains technically sound, organized and relevant. Such a group develops partnerships and gradually becomes a local institution with a growing sense of mission and political voice. But virtually all CBWM groups need the underpinning of a support network and a larger programme. Accessing necessary equipment, managing data and providing the training in existing and new parameters is typically beyond the scope of a local group. A sustainable CBWM programme, which accommodates the diverse needs of several groups, requires strong partnerships and institutional support. Building such a programme has challenges because required activities often do not coincide with traditional programmes of universities or development organizations.

In summary, because data collection and management is an essential part of a successful CBWM programme, information about water monitors, groups, watersheds, sites and water data must be organized for analyses and dissemination in appropriate forms. Support for database development and maintenance must typically be provided by an outside organization. Efforts to provide the technical infrastructure for water monitoring and reporting are currently being spearheaded by the International Center for Aquaculture and Aquatic Environments of Auburn University. The Center is planning for increased interdisciplinary approaches to addressing aquatic resource management, and a worldwide network of CBWM groups, called Global Water Watch (GWW), is being formed. The aim is to provide a way for local CBWM groups to enter water monitoring data in the GWW database, via the website established at www.globalwaterwatch.org. Monitoring groups, policymakers, educators and the general public will then be able to access the information in summarized forms. There are several options for graphing, mapping and statistical analyses that are adapted to group needs. As new groups voluntarily become a part of GWW, and additional funding is secured, the database and website will be improved and expanded.

References

Bellows, B., Buenavista, G. and Ticsay-Rusco, M. (eds) (1995) *Participatory Landscape Lifescape Appraisal*, vol. 1: *The Manupali Watershed, Province of Bukidnon, the Philippines*. SANREM CRSP Philippines: The Practice and the Process. SANREM Research Report No. 2-95. University of Georgia, Athens, Georgia.

Bin, Li (1994) The impact assessment of land use change in the watershed area using remote sensing and GIS: a case study of Manupali watershed, the Philippines. MSc. thesis, Asian Institute of Technology, Bangkok.

Cody, R.P. and Smith, J.K. (1997) *Applied Statistics and the SAS Programming Languages*, 4th edn. Prentice-Hall, Upper Saddle River, New Jersey.

Deutsch, W.G. and Busby A.L. (1994) *Quality Assurance Plan for Chemical Monitoring for Alabama Water Watch*. Auburn University, Auburn, Alabama.

Deutsch, W.G. and Busby A.L. (1999) *Quality Assurance Plan for Bacteriological Monitoring for Alabama Water Watch*. Auburn University, Auburn, Alabama.

Deutsch, W.G. and Orprecio, J.L. (2000) Formation, potential and challenges of a citizen volunteer water quality monitoring group in Mindanao, Philippines. In: Cason, K. (ed.) *Cultivating Community Capital for Sustainable Natural Resource Management. Experiences from the SANREM CRSP*. Sustainable Agriculture and Natural Resource Management Collaborative Research Support Program. Athens, Georgia, pp. 13–20.

Deutsch, W., Busby, A.L., Winter, W., Mullen, M. and Hurley, P. (1998) *Alabama Water Watch: The First Five Years*. Research and Development Series 42, International Center for Aquaculture and Aquatic Environments. Auburn University, Auburn, Alabama.

Deutsch, W.G., Orprecio, J.L., Busby, A.L., Bago-Labis, J.P. and Cequiña, E.Y. (2001a) Community-based Water Quality Monitoring: From Data Collection to Sustainable Management of Water Resources. In: Coxhead, I. and Buenavista, G. (eds) *Seeking Sustainability: Challenges of Agricultural Development and Environmental Management in a Philippine Watershed*. Philippine Council for Agriculture, Forestry and Natural Resources Research and Development, Department of Science and Technology: Los Baños, Laguna, Philippines, pp. 138–160.

Deutsch, W.G., Orprecio, J.L. and Bago-Labis, J.P. (2001b) Community-based water quality monitoring: the Tigbantay Wahig Experience. In: Coxhead, I. and Buenavista, G. (eds) *Seeking Sustainability: Challenges of Agricultural Development and Environmental Management in a Philippine Watershed*. Philippine Council for Agriculture, Forestry and Natural Resources Research and Development, Department of Science and Technology: Los Baños, Laguna, Philippines, pp. 184–197.

Deutsch, W.G., Orprecio, J.L., Busby, A.L., Bago-Labis, J.P. and Cequiña, E.Y. (2005) Community-based hydrologic and water quality assessments in Mindanao, Philippines. In: Bonell, M. and Bruijnzeel, L.A. (eds) *Forests, Water and People in the Humid Tropics*. Cambridge University Press, Cambridge, pp. 134–149.

Foglia, K. (1995) *Choosing a Sustainable Future. SANREM CRSP Annual Report*. Sustainable Agriculture and Nature Resource Management Collaborative Research Support Program. University of Georgia, Athens, Georgia.

Paunlagui, M.M. and Suminguit, V. (2001) Demographic development of Lantapan. In: Coxhead, I. and Buenavista, G. (eds) *Seeking Sustainability: Challenges of Agricultural Development and Environmental Management in a Philippine Watershed*. Philippine Council for Agriculture, Forestry and Natural Resources Research and Development, Department of Science and Technology, Los Baños, Laguna, Philippines, pp. 31–47.

Poudel, D.D. and West, L.T. (1999) Soil development and fertility characteristics of a volcanic slope in Mindanao, the Philippines. *Soil Science Society of America Journal*, 63, 1258–1273.

Rola, A. (2004) Development, policies, institutions and environment in the uplands of Southeast Asia. Paper presented at the International Research Conference on Sustainable Agriculture and Natural Resources Management, 'Land Use Changes in Tropical Watersheds: Causes, Consequences, and Policy Options'. 13–14 January. Quezon City, Philippines.

4 How Do National Markets and Price Policies Affect Land Use at the Forest Margin?

Evidence from the Philippines*

I. Coxhead,[1] A. C. Rola[2] and K. Kim[3]

[1]Department of Agricultural and Applied Economics, University of Wisconsin-Madison, 427 Lorch Street, Madison, WI 53706, USA, e-mail: coxhead@wisc.edu; [2]University of the Philippines at Los Baños, College, Laguna 4031, Philippines, e-mail: arola@laguna.net; [3]Seoul National University, Gwanak Campus Bldg 200, Seoul, Republic of Korea, e-mail: kimk@snu.ac.kr

Introduction

Poor farmers in developing countries are the primary managers of an increasingly scarce natural resource, productive agricultural land. Their decisions, while they may be privately optimal, often conflict with social goals of resource conservation. This is clearly true when farmers intensify production on soils that are easily eroded, and when agricultural expansion takes place through the conversion of forests and other permanent cover to seasonal crops.

The empirical literature on tropical deforestation and land degradation is rich with studies of resource use by households whose actions are constrained by poverty, market failures and risk aversion (e.g. Southgate, 1988; Anderson and Thampapillai, 1990; Shively, 1997). The literature typically locates such immediate motivational factors within a broader context of absence or non-enforcement of property rights (resulting in open-access forest lands and tenure insecurity on farmed lands), and population pressure. These are identified as providing the enabling environment for forest clearing

*This chapter originally appeared as Coxhead et al. (2001). Copyright 2001, used by permission of the authors and The University of Wisconsin Press. Some editorial changes have been made. For data collection we are grateful to Isidra Bagares and the Philippine Bureau of Agricultural Statistics. Financial support was provided by the University of Wisconsin Graduate School, and by USAID through the SANREM CRSP.

©CAB International 2005. *Land Use Change in Tropical Watersheds: Evidence, Causes and Remedies* (eds I. Coxhead and G. Shively)

and unsustainable patterns of agricultural land use by upland farmers (Pingali, 1997; Rola and Coxhead, Chapter 2, this volume). There are also a number of analytical models exploring the influence of broader economic forces like price policies and wage trends on elements of the upland agricultural decision set, such as soil conservation (Barbier, 1990; Barrett, 1991) and deforestation (Angelsen, 1999). At the broadest level are general equilibrium papers in which intersectoral linkages, through factor markets, product markets and trade, are seen to influence upland decisions (Lopez and Niklitschek, 1991; Coxhead and Jayasuriya, 2003; Coxhead and Shively, Chapter 6, this volume).

Looking across all the types of models one finds a wide array of assumptions about the economic links between upland economies and the national economies in which they are located (Angelsen, 1999, in particular, explores many variations). The choice of market assumptions conditions the behaviour of a model and thus the policy conclusions that are drawn from it. As an example, general equilibrium approaches to deforestation, by allowing for intersectoral labour mobility, convey the idea that upland population 'pressure' is a response to economic incentives, rather than an exogenous determinant of actions as in some of the other models.

Similarly, the assumptions commonly made about product markets vary greatly. Given the importance of a correct specification of market structure and pricing, there are surprisingly few studies that bring empirical evidence to bear on the market and policy aspects of upland agricultural resource use decisions. The goal of this chapter is to encourage a move in that direction, as a complement to the existing body of household-level analyses and land use response studies (such as Coxhead and Demeke, Chapter 5, this volume).

It is our thesis that the design of upland projects directed at influencing smallholders' land conversion and land use decisions in the direction of 'sustainability' could be greatly improved by better integrating information on market- and sector-level incentives with information on household-level decisions and constraints. Perhaps because of a lack of data and empirical analysis, project solutions to deforestation and agricultural land degradation in developing countries focus mainly on direct interventions through technology transfer, institutional innovations and other household-level actions. The role of policy (and especially its less direct manifestations through intersectoral product and factor markets), is generally given little emphasis.[1] The obverse of this problem is a general neglect at the policy level of the intersectoral and environmental impacts of trade and agricultural pricing policies.[2] Both forms of myopia may have restricted the domain of possible project solutions to upland environmental problems, and indeed may have increased the probability that projects will fail because of conflicting messages contained in the direct and intersectoral signals from economic policies.

To illustrate this point, consider an upland economy producing two goods, one via a land-intensive technology and the other via a labour-intensive technology.[3] Define the relative price of the latter to the former as P, and price of a single lowland good (relative to the upland land-intensive good) by Q. There are thus three product markets: two for upland goods and one for an aggregate lowland good. Assume that upland production uses land and labour, and that in order to be brought into production, land must first be cleared of forest, an activity that uses labour. Profit-maximizing upland producers will thus allocate labour to forest clearing and farming, and to one crop or the other, in response to P. If there are links to the lowland economy, then Q will also play a role in these decisions.

Suppose first that all three goods are freely traded with the rest of the world at given world prices, and that the world price of Q increases. Whether this change has any effect in uplands, and if so in what manner, will depend on interregional labour markets. If labour is immobile between regions, the increase in Q will have no effect. If it is mobile, the increase will raise labour productivity in lowlands relative to uplands and induce migration out of the uplands. Since labour is needed for forest clearing as well as farming, deforestation must decline. However, now suppose that the labour-intensive good in uplands is not traded, so its price depends on domestic demand and supply; thus $P = P(Q)$. Now an increase in Q will have different effects on uplands depending on whether the goods are substitutes in consumption as well as supply side factors and income effects.

Similarly, imagine that Q is constant but that more productive new technologies are adopted by producers of one upland crop. If products are freely traded at world prices, upland labour productivity will rise and both the share of land and the total area planted to the crop experiencing technical progress will increase; deforestation will rise in this case. On the other hand, if demand for the crop is downward-sloping (whether due to local or national non-tradability), then technical progress will alter P and the total area of upland, as well as the share planted to the labour-intensive crop, could rise or fall.

These simple illustrations highlight the sensitivity of deforestation and upland land use outcomes to the conditions of agricultural markets. Under one set of assumptions, technical progress in an upland crop is predicted to increase deforestation; under another, deforestation could fall. Conversely, a national policy innovation that alters P or Q (or both) has the potential to induce changes in deforestation and land use even when the policy measure is not directly related to agriculture. This is so even when all goods' prices are exogenous, if labour is mobile between regions. Lastly, when upland farmers are risk-averse, the entire argument can be restated (with modifications as appropriate) using price variances as well as levels.

In the rest of the chapter we focus on a Philippine case study. We first provide a brief survey of major macroeconomic and policy trends and their possible effects on upland resource use decisions. While we have information about macroeconomic and economy-wide phenomena, and about upland farmers' decision making processes, we know little about the nature and strength of market links between the two. We then use econometric analysis to examine linkages between national and farm-gate prices on the basis of recent data from the Manupali River watershed.

Growth, Policies and Upland Resource Use in the Philippines

After a long period of stagnation, the pace of aggregate economic growth in the Philippines accelerated in the 1990s. However, the degree of dependence on agriculture and natural resources remains high by South-east Asian regional standards. This is a function of earlier decades of slow growth and rapid population increase, which maintained a high level of dependence on agriculture. It can thus be argued that the persistence of pressure on forest and upland agricultural land resources is in part a consequence of poor macroeconomic performance.[4]

In the early post-war years migration to heavily forested frontier areas in the Philippines was officially encouraged as a means of alleviating economic and political pressures generated by increasing population and stagnating technology in the country's rice-growing heartlands (Rola and Coxhead, Chapter 2, this volume). In subsequent decades, continued spontaneous internal migration has been fostered by low rates of non-agricultural labour absorption, as well as a series of labour-saving technical changes in lowland irrigated agriculture (Jayasuriya and Shand, 1986), in the face of sustained high rates of overall labour force growth. The resulting increases in landlessness and unemployment stimulated searches for open-access resources from which incomes, however tenuous, could be earned. The outcome was a trebling of upland population between 1950 and 1985, from 5.8 million to 17.5 million, and annual growth rates of upland *cropped* area of greater than 7% over the same period (Cruz *et al.*, 1992). The evidence that macroeconomic instability and growth (or the lack of it) in non-agricultural sectors were major forces driving migration and upland land use decisions is compelling, if circumstantial (Cruz and Repetto, 1992).

There is a strong suggestion that microeconomic and trade policies also promoted forest conversion and intensification in upland agriculture. In commercially oriented upland agriculture – or even simply where labour is mobile into or out of upland areas – agricultural price policy can exert a significant, though not immediately observable, influence on natural resource management. In the Philippines there is evidence of a pervasive policy bias in favour of crops – such as maize and temperate vegetables – whose cultivation is most strongly associated with upland agricultural land degradation, soil erosion and related water pollution. This commodity bias emanates mainly from national-level economic policies, some of them unrelated to agriculture; it has been complemented in the past by the allocation of agricultural research resources; and it appears not to have been offset by economic policy measures in favour of more environmentally friendly cultivation techniques.

Throughout the post-war era successive Philippine governments have pursued self-sufficiency in grains, along with cheap consumer cereals prices, as key components of food security and income redistribution strategies. Philippine cereal yields are low by Asian standards, and with relatively little spending on agricultural infrastructure and technology, yields have not risen as quickly as in comparable countries. Consequently, grain output growth in

uplands has been due primarily to area expansion. Given the political importance of self-sufficiency, grain imports are tightly circumscribed, and this in
turn has maintained domestic producer prices at levels well above the domes-
tic-currency equivalents of world prices.[5]

Vegetable production has also received substantial policy support. Import
bans imposed in 1950 on fresh potato, cabbage and other horticultural crops
(and reiterated in legislation as recently as 1993) were repealed and replaced
by tariffs only in 1996 (see below). Demand for these foods grows with *per
capita* income and urbanization. Since supply growth is limited by trade restrictions and climatic constraints, their prices have tended to rise more rapidly
than the general price level, and certainly more rapidly than prices of most
exportable crops and staple grains. For potato, the ban raises Philippine *farm
gate* prices to nearly double the imputed c.i.f. (landed) *wholesale* price of
imports, if they were permitted (Coxhead, 1997).

The Agricultural Tariffication Act of 1996 brought Philippine agricultural
policy into compliance with the Uruguay Round of the GATT. Quantitative
import restrictions on maize and vegetables were replaced by tariffs, and minimum access volumes (MAVs) were specified for each product. (Under WTO
rules, the MAV is the volume of a product that is allowed to be imported at a
lower rate of duty than the maximum bound rate.) For the period to 2004, in-
quota maize tariffs (those applying to MAV imports, which themselves cover
roughly 50% of annual imports) remain at 35%. Out-quota tariff rates for
maize, set at 100% in 1996 have fallen little in subsequent years (similar
changes apply to vegetables). These reforms, although they constitute important steps in the direction of more open trade, ensure that upland farmers will
continue to benefit from protection at significantly higher rates than most
other sectors for the foreseeable future.

Trade and price policy biases are also reflected in the allocation of agricultural research funds. Most important among these for uplands are maize
programmes. A number of provinces (including Bukidnon, from which our primary data are drawn) were designated as 'key production areas (KPAs)' for
maize in the Philippine government's Grain Production Enhancement
Program (GPEP). Farmers in KPA areas are eligible for subsidies and supports
directed at increasing maize production, and are the first beneficiaries of
research and development directed at increasing maize yields (Philippine
Department of Agriculture, 1994). Similarly, temperate climate vegetable
crops are also the targets of disproportionate research resource allocations
(Coxhead, 1997).

This brief review of Philippine growth strategy and policies has indicated a
number of channels through which decisions concerning use of upland forest
and farmland are likely to be influenced. In the longer term, a successful development strategy would have raised lowland and non-farm labour productivity
faster than in uplands and diminished the economy's susceptibility to destabiliz-
ing macroeconomic shocks; all these should have reduced net migration to
uplands and by extension, reduced pressures on forest and land resources. Trade
policy liberalization would in general have promoted growth of export-oriented

non-agricultural sectors and might have preserved the profitability of some upland perennial export crops, such as coffee, relative to annual crops, and this in turn might have caused some redirection of input subsidy schemes and R&D resources away from import-competing crops and towards more promising sectors. Moreover, this review makes it clear that there are many potential policy changes at the macroeconomic level or in trade and agriculture sector policy that could affect upland resource allocation. On the basis of this evidence any project directed at influencing upland resource allocation towards a 'sustainable' path should at least be cognizant of this broader setting, if not actively involved in trying to alter it.

The evidence we have reviewed, however, is strictly circumstantial. Questions remain as to the strength and nature of linkages between uplands and the national economy, and it is in this field of inquiry, as previously noted, that specific data and evidence are lacking. The gap creates room for competing hypotheses about the upland economy, and these in turn imply different diagnoses of upland environmental problems and their solutions. In the next part of the chapter we describe the site from which primary data have been drawn in an attempt to fill this gap.

The study site

The research site is in Lantapan municipality, Bukidnon province, the geography of which was described in Chapter 1. Most households in Lantapan are poor by Philippine standards, and rainfed agriculture dominates the local economy. Low-lying flatlands are devoted to rice and sugarcane and maize–sugarcane systems dominate rolling mid-altitude areas. At higher elevations maize is the predominant crop, planted alongside coffee and temperate-climate crops – beans, tomatoes, cabbages and potatoes. The latter two crops require cool nighttime temperatures and so are generally grown above 1000 m in fields adjacent to and even within the boundary of Mt Kitanglad Range Nature Park, which forms the northern boundary of the municipality.

In both spontaneous migration and official programmes since the 1950s, Bukidnon was a major destination, and watersheds like the Upper Manupali were choice locations. Population growth rates peaked at 10% per year in the 1950s, with most of the increase due to in-migration from economically depressed areas of the central and northern Philippines (NSO, 1990). In the decade from 1970 Lantapan's population increased at an average annual rate of 4.6% (NSO, 1990). Since 1980 the annual population growth rate has averaged 4%, far higher than the Philippine average of 2.4%.

Commercial agriculture has expanded along with population. Internal migrants introduced commercial cultivation of potato, cabbage and other vegetables in the 1950s. More recently infrastructural improvements, coupled with increasing demand for vegetables and feed (yellow) maize, have ensured that commercial agriculture in the province continues to adapt and thrive. Maize and vegetable production have flourished; falling transport costs have helped these become primarily commercial crops, exported to the national economy where formerly they had been little traded outside the province.

Is there an environmental problem?

In Lantapan, agricultural expansion has occurred substantially at the expense of perennial crops, including forest (see Table 1.2). Other things being equal, the replacement of perennial land uses with short-season and annual crops on sloping lands is associated with rapid increases in soil erosion and land degradation. Field measurements and experiments with the cultivation of maize and vegetables under a range of management regimes in Lantapan confirm rapid erosion and soil nutrient and organic matter depletion (Midmore *et al.*, 1997). In spite of these negative effects of the spread of annual crops, few farmers display deep knowledge of soil degradation relationships (Midmore *et al.*, 1997). Land fallowing and crop rotation is rare and usually undertaken only when yields decline to the point of economic losses in the current season. Although soil erosion and land degradation problems appear to be widespread, few farmers report significant investments in soil-conserving structures or technologies. Failure to adopt soil conservation measures is correlated with tenure insecurity; fields held in private title are more likely to be fallowed regularly, to have tree crops, and to have perennial grasses planted on boundaries than are fields operated by tenants. Farmers are more likely to practice contour ploughing on owned than on rented land (Coxhead, 1995; Midmore *et al.*, 1997).

In upland watersheds, agricultural intensification without adequate soil management has deleterious effects on-site and off-site. Intensive cultivation of annual crops in general, and the increased use of fertilizer, pesticides and other chemicals on vegetable crops in particular, are likely to degrade water quality and may create health problems for farm families and those living downstream. Lantapan-based water quality monitoring reveals both qualitative and quantitative evidence of such problems. Perceptions of pesticide residues have made some residents reluctant to water animals in streams during or after heavy rain. Measures of total suspended solids (TSS) across subwatersheds are considerably higher where agricultural cultivation is more widespread, in spite of lower average slope, and seasonal TSS peaks coincide with months of intensive land preparation activity. Many of the more noticeable changes in water quality and seasonal flows have occurred 'well within human memory' (Deutsch *et al.*, 1998, p.12).

Finally, the unchecked expansion of agriculture into the national park poses a potential threat to the biological integrity of the remaining forest. In the early post-war years forest encroachment was driven mainly by commercial logging, but in the past two decades the expansion of small maize and vegetable farms has been the primary impetus, with decisive contributions from road development and the lack of established property rights in land (Cairns, 1995). Concerns arising from forest removal and degradation include such specific phenomena as loss of watershed function (especially with clearing in the headwaters of creeks), changes in the quantity and seasonal distribution of water flow in springs and rivers, loss of wildlife habitat, and reduced availability of forest-based foods and raw materials – as well as more general, and less easily quantified, phenomena such as biodiversity loss.

In summary, evidence on environmental problems in the watershed provides emphatic support for two arguments. First, the natural resource base of the watershed is undergoing degradation of a nature and at a rate without modern precedent. Second, much if not most of the degradation can be attributed directly or indirectly to the spread of intensive agricultural systems based on maize and vegetables. In this setting, the designers of forest conservation and sustainable agriculture projects debate both the root causes of deforestation and land degradation, and the means by which they should be addressed.

Analysis of Markets, Prices and Land Use Decisions

Our research focuses on factors influencing land use in the middle and upper-watershed areas, on relatively steep and easily eroded valley sides and at the forest margin. The major crops grown are maize (both for feed and for human consumption) and vegetables – especially cabbage, beans and potato. In the analysis that follows we concentrate on maize, as by far the most important crop, in terms both of land use and of net farm incomes, within the study site. Nationally, too, the area planted to maize is second only to rice, and maize accounts for by far the greatest part of upland agricultural land use.

An initial survey of the Lantapan site had characterized agriculture in the upper watershed as 'subsistence' or 'semi-subsistence' (Bellows, 1993). However, our data reveal clear commercial motivations for almost all farmers.[6] More than 50% of maize production is destined for market, and vegetable crops such as cabbage, potato and beans strictly for sale, with home consumption accounting for less than 10% of production in each case (Coxhead, 1995).

Econometric analysis of land use decisions by upland farmers in this and in comparable Philippine locations indicates that their land allocations respond to relative prices, and to price variability, in statistically significant ways (Shively, 1998; Coxhead *et al.*, 2002; Coxhead and Demeke, Chapter 5, this volume). A question remains as to the relative importance of *markets,* as well as of national policies operating through them, as conditioning influences over farmers' decisions. If farm gate prices or their variability are important determinants of land use decisions, what are the determinants of farm gate prices?

Market integration and price causation

As argued earlier, understanding the nature of market links between uplands and the rest of the economy is critical to the efficiency of project and policy design. If markets within the study site were isolated from or only weakly associated with regional markets (the 'semi-subsistence' hypothesis), we would expect to see seasonal or even longer-term divergence between trends in local and regional prices. Further, we would be unable to see evidence that local prices are driven by national prices.

The tests of market integration and the direction of causation are important for both economic and environmental reasons. Under current production technologies maize, potato, cabbage and other intensive crops in Lantapan generate annual erosion and soil nutrient losses far in excess of natural regeneration rates. Remoteness and poor quality of infrastructure are frequently taken to indicate that market links to the rest of the economy are tenuous at best. This, if true, would have two important implications for policy and project design. It would mean that agricultural prices and trade policies – standard instruments for influencing agricultural resource allocation in lowlands – could be expected to have little or no effect in uplands. By extension, the most effective instruments for promoting sustainable agriculture in uplands would be interventions such as technology transfer, extension and education.

Alternatively, if markets are integrated but farm-gate prices are most significantly influenced by local production, then supply and price in upland agriculture will tend to move in opposite directions. If an increase in local supply drives down prices, then the profit-maximizing level of local output will be lower than if prices had not been affected. In this case the price-reducing effects of local adoption of supply-increasing innovations such as new technologies or more efficient management practices might be expected to act as a 'natural brake' on the expansion of agriculture at the forest margin. However, these effects (even if they were observed) are likely to obtain only in the short run, since integration with the larger market will probably neutralize local effects in the longer run.

Unfortunately, there is no widely accepted statistical test of market integration, only tests of relationships between prices that (if confirmed) can be said to be consistent with integration. Theory tells us that if two markets are linked through trade, then under normal circumstances, differences in prices net of margins between the two markets create opportunities for arbitrage. Goods will flow between the two markets – trade will occur – until the price difference is eliminated. Statistically, if the prices in the tested markets are non-stationary (i.e. that they are trending over time rather than merely following a random walk) then the markets are integrated if their price series are cointegrated, meaning that there is a (single) stationary long-term relationship between them.

We investigated the time series properties of the price data on yellow maize, white maize, potatoes and cabbage in Lantapan and the main regional wholesale market and in each case found the series to be stationary.[7] Therefore, we cannot conduct a *statistical* test of long-term market integration. However, our observation confirms that trade between farms and the wholesale market is regular, seasonally consistent, and consists of high volumes; a statistical finding of no integration would be a very great surprise. Studies using aggregate data have indicated clearly that Philippine grain markets are integrated across regions and provinces (Silvapulle and Jayasuriya, 1994; Mendoza and Rosegrant, 1995).

Examining the short-run dynamics of the price series permits tests of the hypotheses that upland farmers are price-takers and that national market and policy signals affect local prices. Our econometric method proceeds as follows. We fit the data to a set of regression equations, each of which has the price of a crop in one market as the dependent variable, and its own lagged values, as well as the current and lagged values of the prices of the same crop in other markets, as explanatory variables. Hypothesis tests on the coefficient estimates of these equations provide information about the direction of causation. As an example, for two markets A and B, when a price change in market A is shown to precede price changes in market B, we describe the price in A as 'Granger-causing' that in B. In our study, confirmation that the local price Granger-causes the regional price would provide support for the 'natural brake' idea referred to above, that expanded production of a crop within the watershed will cause its price to fall, at least in the short run. Conversely, confirmation that the regional price causes the Lantapan price would indicate a need to focus on agricultural price and trade policies as longer-term influences over farmers' land use and crop production decisions.[8]

The test of causation is also a test of a sufficient condition for short-term market integration, so long as at least one causal relationship is confirmed. It is, however, important to note that strictly speaking, our method provides what is best described as circumstantial evidence on integration and causation. The conclusion of 'causation' is reached by observing temporal precedence, but no specific economic mechanism of causation can be identified.

We apply these tests to weekly maize, potato and cabbage prices in Lantapan and the main regional market. Crop price data were collected weekly from traders at several points in Lantapan, from provincial centres and from the main regional wholesale market ('Agora') in Cagayan de Oro, the regional capital and port. Much of the produce sold in the Agora market is shipped directly to Manila, the national capital and central market, either for processing or for sale; accordingly, Agora prices track the benchmark Manila prices. In this analysis we concentrate on the Lantapan–Agora market relationship (the data series are summarized in Figs 4.1–4.3).

To account for the time series properties of the data we employ a vector auto-regression (VAR) model (Sims, 1980). The VAR approach to time series analysis is controversial. As Cooley and LeRoy (1985) have pointed out, the VAR is 'atheoretical' in the sense that it embodies no explicit economic theory. However, when restrictions in the VAR model, in terms of choices of variables and lag lengths, are weaker than the restrictions imposed on structural models, the VAR approach can provide a foundation for testing hypotheses based on *a priori* reasoning (Backus, 1986). In our investigation of price relationships, we use both economic and econometric tools to choose variables and lag lengths. We thus view the VAR approach as a complement to the structural models implied by theory. Specifically in the case of Lantapan, the quality of transport infrastructure, high frequency of public and private travel, and the distance (130 km, or at most 5 h) to the major market all suggest that price signals can be exchanged, and arbitrage occur, well within the 2-week interval implied by a two-period lag structure.

The structural equations of the VAR model (with two-period lags, suppressing crop-specific subscripts) are:

$$PL_t = \alpha_1 PA_t + \beta_{11} PL_{t-1} + \beta_{12} PA_{t-1} + \gamma_{11} PL_{t-2} + \gamma_{12} PA_{t-2} + v_{PLt}$$

$$PA_t = \alpha_2 PL_t + \beta_{21} PL_{t-1} + \beta_{22} PA_{t-1} + \gamma_{21} PL_{t-2} + \gamma_{22} PA_{t-2} + v_{PAt}$$

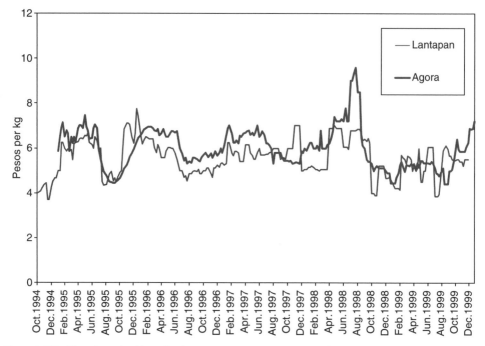

Fig. 4.1. Weekly price of yellow (feed) maize, 10/94–12/99 (pesos/kg).

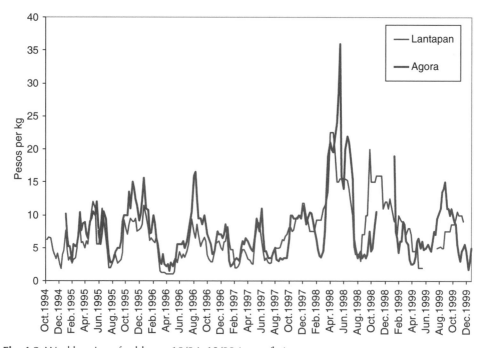

Fig. 4.2. Weekly price of cabbage, 10/94–12/99 (pesos/kg).

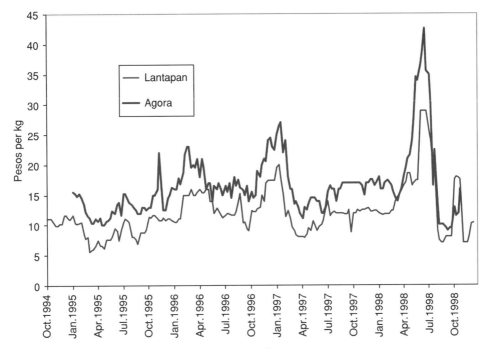

Fig. 4.3. Weekly price of potato, 10/94–12/98 (pesos/kg).

where PL_t and PA_t are prices in Lantapan and in the Agora regional market respectively, and v_{PLt} and v_{PAt} are error terms that we assumed are serially and mutually uncorrelated. Eliminating current-period variables from the right-hand sides of these equations yields a reduced form:

$$PL_t = \varphi_{11}PL_{t-1} + \varphi_{12}PA_{t-1} + \varphi_{13}PL_{t-2} + \varphi_{14}PA_{t-2} + \varepsilon_{1t}$$

$$PA_t = \varphi_{21}PA_{t-1} + \varphi_{22}PL_{t-1} + \varphi_{23}PA_{t-2} + \varphi_{24}PL_{t-2} + \varepsilon_{2t}$$

in which ε_{1t} and ε_{2t} are unobservable variables which are the serially uncorrelated innovations in the PL and PA processes.

Granger causality tests utilize test statistics computed from the VARs. A variable m_t is said to *fail to Granger-cause* another variable y_t relative to an information set consisting of past values of m_t and y_t if

$$\hat{E}[y_t \mid y_{t-1}, m_{t-1}, y_{t-2}, m_{t-2}, \ldots] = \hat{E}[y_t \mid y_{t-1}, y_{t-2}, \ldots]$$

where $\hat{E}$ denotes a linear projection of the dependent variable. In our example, this means that PA does not Granger-cause PL relative to an information set consisting of past values of PA and PL if (and only if) the estimates of φ_{12} and φ_{14} are equal to zero. In practice, an F-test can be used to test the null that one variable does not Granger-cause another.

The results of these F-tests are summarized in Table 4.1. All markets display some form of causation, and so we conclude that local and regional markets are integrated for all crops in the study. For *yellow maize* and *white*

Table 4.1. Summary of results of Granger causality tests for maize and vegetable prices.

Crop	Test[a]	R^2	DW[b]	$F(N; df)$	P value[c]	Comments
Weekly data						
Yellow maize	Agora → Lantapan	0.75	1.97	3.22 (182; 2,176)	0.042	One-way causation
	Lantapan → Agora	0.86	2.04	0.91 (182; 2,176)	0.403	
White maize	Agora → Lantapan	0.89	1.95	8.25 (162; 2,156)	0.004	One-way causation
	Lantapan → Agora	0.95	1.96	0.39 (162; 2,156)	0.680	
Avg. potato	Agora → Lantapan	0.81	1.95	6.61 (157; 2,151)	0.002	Two-way causation
	Lantapan → Agora	0.84	2.08	7.17 (157; 2,151)	0.001	
Cabbage	Agora → Lantapan	0.86	1.97	2.88 (170; 2,164)	0.005	Two-way causation
	Lantapan → Agora	0.68	1.96	5.60 (170; 2,164)	0.004	
Monthly data						
Avg. potato[d]	Agora → Lantapan	0.75	2.05	13.80 (83; 2,76)	0.001	One-way causation
	Lantapan → Agora	0.83	2.12	0.77 (83; 2,76)	0.470	
Cabbage	Agora → Lantapan	0.61	1.90	3.36 (41; 2,35)	0.046	One-way causation
	Lantapan → Agora	0.56	1.99	0.34 (41; 2,35)	0.710	

[a] Arrows indicate the direction of causation being tested, so for example 'Agora → Lantapan' indicates a test that Agora price Granger causes Lantapan price.
[b] Durbin–Watson statistic.
[c] $P < 0.01$ indicates rejection of the null hypothesis (no causation) at 1% significance level; $0.01 < P < 0.05$ indicates rejection at 5%; $0.05 < P < 0.1$ indicates rejection at 10%.
[d] Biweekly data for average prices of large and medium potatoes.

maize, the direction of causation runs from wholesale market to farm gate. Maize prices in the watershed are driven entirely by prices in provincial and national markets. For *potato*, weekly data indicate two-way causation: farmgate prices are influenced by wholesale prices, but a local supply shock in Lantapan may also have a short-run effect in wholesale markets. Using biweekly data, however, we find a strong one-way relationship between Lantapan and Agora prices, with causality running from the latter to the former. For *cabbage*, the weekly data show a strong influence of Lantapan prices on wholesale prices, but monthly data show that when very short-term fluctuations are smoothed out, cabbage prices are determined in the regional market and not within the watershed.

To summarize, our results indicate that markets for the major crops grown in the watershed are integrated in the short term with broader regional markets. They also provide strong evidence for all crops that an expansion of supply within the watershed will have no measurable influence on its prices in wholesale markets, beyond a period of one or two weeks for vegetable crops. Rather, the evidence is that farmers in the watershed are price takers in regional and national markets.

Market and policy linkages in Lantapan

If markets are integrated as we have argued, and given that short-term causality runs only from regional to local market, what can we conclude about the implications of national policies for upland land use in Lantapan?

For the reasons indicated earlier, we cannot as yet quantify the effects of changes in the trade policy regimes that underpin domestic market conditions for both maize and vegetables. For vegetables, import bans that prevailed until 1996 have been replaced with tariffs at prohibitive rates; in effect, there has been no trade policy change. For maize, in spite of the shift from quantitative restrictions to the MAV system with tariffs after 1996, announced trade policy changes are being introduced very gradually and in practice have changed very little. However, our finding that upland farmers are price-takers in regional markets makes it clear that any meaningful policy changes, were they to occur, would have direct effects on farm-gate prices in the uplands.

Of potentially greater interest is the observation that revenue instability, the phenomenon that risk-averse farmers strive to avoid, has intersectoral as well as local sources, even in a market (such as maize) which is subject to price stabilization. Our data span the economic crisis that engulfed South-east Asian countries in the late 1990s, beginning when the Thai currency collapsed in July 1997. While the crisis took different forms in each affected economy, there were three elements common to all. There was a sharp drop in overall economic growth, and there were sudden, unexpected and repeated re-evaluations of exchange rates that had previously been effectively pegged to the US dollar. As a result there was a big increase in uncertainty among producers within the affected countries about final demand and prices, input prices, and even availability of key inputs such as credit. Since trade policy renders Philippine maize prices largely independent of world prices in the short run, were upland markets affected by the macroeconomic instability reflected in the exchange rate?

We used information about exchange rate variability to define the end-points of the Philippine economic crisis.[9] During the period August 1997 to November 1998, the daily peso–dollar rate fluctuated wildly, whereas before and after this episode, the mean daily change was a fraction of 1% (Fig. 4.4). We use this criterion to divide our data into 'pre-crisis', 'crisis' and 'post-crisis' periods; as Table 4.2 shows, the price variance of yellow maize, the major crop in Lantapan, increased substantially during the crisis, even through the mean price did not. We are then able to make a preliminary identification of the effects of macroeconomic instability on the relationship between farm gate prices and those in national markets. We do this by calculating *impulse response functions*, which record the dynamic response of one data series to a one-time shock ('impulse') in another (see Greene, 1993). For example, the dynamic response of a shock in Agora on the Lantapan price can be captured by $\partial PL_{t+j}/\partial v_{PAt}$.[10] The impulse response measures are thus computed from the same VAR model used earlier to test market relationships, only with the data divided into sub-periods as noted.

The dynamic response of Lantapan maize prices to a shock in the Agora regional market price is shown for a two-period lag model in Fig. 4.5. The 'impulse' is a one-peso per kilogram price shock, so the figures on the vertical axis of the graph are pesos per kilogram in the Lantapan market (the mean pre-shock maize price was about 6 pesos/kg, so this represents a shock of about 16%). In the pre-crisis and post-crisis periods, a shock in the Agora price yields a maximum rise in local prices of about 3% (0.2 pesos). The impulse response peaks 3 weeks after the shock and drops very sharply to a negligible amount by the 5th week after the shock. During the crisis, the peak is much larger (6%), and is sustained over several months.

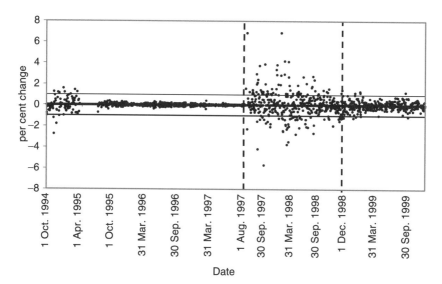

Fig. 4.4. Daily fluctuations of the Philippine peso against the US dollar.

Table 4.2. Moments of yellow maize prices before, during and after the exchange rate crisis.

Item	Pre-crisis	Crisis	Post-crisis
Lantapan mean price (pesos/kg)	5.57	5.85	5.19
Agora mean price (pesos/kg)	6.15	6.25	5.23
Lantapan price variance	0.627	0.774	0.462
Agora price variance	0.526	1.277	0.221
Exchange rate (peso/US$)	25.9	38.7	38.9
Exchange rate variance	0.468	16.892	0.817

Note: Periods are as defined in text and illustrated in Fig. 4.4.

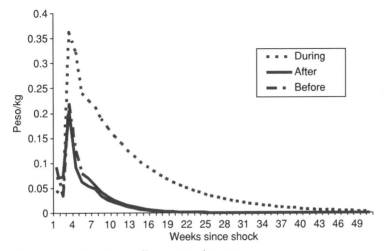

Fig. 4.5. Impulse response function: yellow maize (lag = 2).

Comparing responses during the crisis period and in the earlier and later periods, we see that in the post-crisis era the signal from the leading Agora price to the Lantapan price is very much more 'noisy' than in the prior period. The price dynamics indicate that during the crisis, a temporary disturbance in the Agora series induces a larger and longer-lived response in farm-gate prices.

The impulse response analysis suggests that the effects of macroeconomic instability find their way into the behaviour of prices that guide farming decisions even in areas far from the main regions and sectors of economic activity. The economic signals upon which upland farmers make resource allocation decisions are not independent of conditions in national markets and in the macroeconomy.

Conclusions

Commodity market development, along with policy biases, has contributed to deforestation and the adoption and spread of relatively erosive crops, produced using relatively land-degrading technologies, in the upland Philippine watershed of our study. The environmental ill-effects of these crops could be minimized by adoption of appropriate technologies, for example to reduce erosion and preserve soil quality. However, only a few farmers in the study site have adopted effective soil conservation measures, and while this is clearly related to tenure insecurity, there is also evidence that among all farmers, the choice of annual commercial crops, and the failure to adopt soil-conserving technologies, have economic as well as institutional roots. If market-driven incentives dominate in farmers' decisions, there is a case for broadening the range of policy instruments brought to bear on the upland environmental problem; moreover, project design may be improved by a different balance of local action and national-level information dissemination and policy advocacy.

We have demonstrated that in spite of remoteness, the farmers in our study area produce for markets that are integrated in the national system. Supply shocks from the site have no effect on prices in broader markets: farmers are price-takers in these markets. National markets transmit both price information and the effects of macroeconomic instability.

While empirical tests of the effects of trade policies on prices await substantive policy changes, it is nevertheless clear that agricultural markets convey the effects of trade policies to the farm gate, even in upland agriculture. Trade liberalization can therefore be expected to reduce the farm-gate prices of maize and vegetables, the two most environmentally damaging crops currently grown in Lantapan and many similar Philippine watersheds.

Finally, anecdotal evidence of the importance of macroeconomic trends in driving upland migration and land use patterns is provided with some additional contemporary support by our finding that the stability of market price relationships is a function of price stability in the overall Philippine economy. During the economic crisis, instability at the macroeconomic level (as reflected in daily exchange rate movements) was associated with a noisier signal from wholesale to farm-gate prices. Future research on the links between deforestation and agricultural expansion should benefit from this exposure of the importance of markets and prices in a typical frontier area of a tropical developing country. A combination of project-specific and more general policy measures is called for if the former are to succeed in changing farmers' actions, and if the latter are not to discourage environmentally sustainable strategies. At a policy level this research, if supported by counterpart studies from other sites, should provoke a reconsideration – and indeed a substantial broadening – of the set of policy instruments available to influence upland agricultural and forest land allocations.

Notes

[1] In the Philippine case, a recent set of national government guidelines for watershed management (PCARRD, 1999) makes only incidental reference to markets as influences on farmer behaviour, and none to policies other than those which have direct effects on land use – zoning, tenure laws and similar.

[2] The following passage from a former undersecretary for policy and planning in the Philippine Department of Agriculture illustrates:

> Policymakers in the Philippines tend to examine economic problems from the perspective of individual consumers and firms, and thus, generate and propose actions and measures focused on directly supporting these entities. In no way have economic policies been evaluated on the basis of their environmental impacts. In rare cases, farmer interests are accounted for. For instance, price controls [on rice and corn] were defended on the basis of their effects on the consumers of staple commodities and the costs of raw materials to enterprises. Rarely were the adverse effects on supply responses as well as the welfare of producer – particularly of farmers and fishermen – considered.
>
> (Tolentino, 1995).

[3] For formal developments of the model in this section, see Lopez and Niklitschek (1991); Deacon (1995); and Coxhead and Jayasuriya (2003).

[4] In Thailand rapid economic growth, and especially the expansion of labour-intensive manufacturing industries, was the major contributor to the stabilizing of agricultural land area during the 'boom' years 1986–96, through out-migration from marginal upland and rural areas (Coxhead and Jiraporn, 1999).

[5] The nominal protective rate (NPR, a measure of the amount by which domestic prices exceed landed import prices) for maize has generally been much higher than for any other major agricultural product, especially after the mid-1970s when maize self-sufficiency was made a policy goal. The NPR averaged 18% in 1970–74, but rose to 42% by 1983–86, and to 62% by 1990 (Coxhead, 2000); it has since remained in this range.

[6] Data on production, input use, land use and sales for major crops were collected annually from a sample of 120 farms in four rounds between 1994 and 1998 (for full details see Coxhead, 1995 and Rola and Coxhead, 1997).

[7] The test for stationarity is conducted with a Dickey-Fuller test of the null hypothesis that each price exhibits a unit root. For example, under an AR(2) representation of yellow maize prices (including seasonal dummy variables), the ADF test statistics for this hypothesis are −4.818 for Agora and −5.307 for Lantapan. At the 5% test level, the critical value for the test is −2.88, so we reject the null hypothesis. We obtain similar results for the other products, which are robust with respect to different lag specifications. We conclude that these price series are stationary.

[8] Both Granger-causality and the test of transmission of shocks (impulse response function) are founded on the vector autoregression (VAR) specification of a price series.

[9] Although the values of the exchange rate are of direct interest in their own right, here we are using exchange rate fluctuations as a proxy for a more general set of macroeconomic conditions. In an open economy, exchange rate depreciation (as occurred during the early part of the crisis) serves as a proxy for (unobservable) inflationary expectations; exchange rate variability is then a proxy for general price instability.

[10] This algebraic derivation involves successive substitution (Greene, 1993).

References

Anderson, J.R. and Thampapillai, J. (1990) Soil conservation in developing countries: project and policy intervention. Agriculture and Rural Development Department, Policy and Research Series, Paper No. 8. The World Bank, Washington, DC.

Angelsen, A. (1999) Agricultural expansion and deforestation: modelling the impact of population, market forces and property rights, *Journal of Development Economics* 58, 185–218.

Backus, D. (1986) The Canadian–U.S. exchange rate: evidence from a vector autoregression. *Review of Economics and Statistics* 68, 628–637.

Barbier, E.B. (1990) The farm-level economics of soil conservation: the uplands of Java. *Land Economics* 66, 198–211.

Barrett, S. (1991) Optimal soil conservation and the reform of agricultural pricing policies. *Journal of Development Economics* 36, 167–187.

Bellows, B. (ed.) (1993) A participatory landscape-lifescape analysis of the Manupali watershed in Bukidnon, Philippines: Characterization of the landscape and identification of research priorities for the SANREM CRSP. Mimeo, University of Georgia, Athens, Georgia.

Cairns, M. (1995) Ancestral domain and national park protection: mutually supportive paradigms? A case study of the Mt. Kitanglad Range National Park, Bukidnon, Philippines. Paper presented at a workshop on Buffer Zone Management and Agroforestry, Central Mindanao University, Musuan, Philippines, August. Mimeo.

Cooley, T.F. and LeRoy S.F. (1985) A theoretical macroeconomics: a critique. *Journal of Monetary Economics* 16, 283–308.

Coxhead, I. (1995) The agricultural economy of Lantapan municipality, Bukidnon, Philippines: results of a baseline survey. SANREM Social Science Group Working Paper No. 95/1. Mimeo, University of Wisconsin, Madison, Wisconsin.

Coxhead, I. (1997) Induced innovation and land degradation in developing country agricul-

ture. *Australian Journal of Agricultural and Resource Economics* 43, 305–332.

Coxhead, I. (2000) The consequences of Philippine food self-sufficiency policies for economic welfare and agricultural land degradation. *World Development* 28, 111–128.

Coxhead, I. and Jiraporn, P. (1999) Economic boom, financial bust, and the decline of Thai agriculture: was growth in the 1990s too fast?, *Chulalongkorn Journal of Economics* 11, 76–96.

Coxhead, I. and Jayasuriya, S.K. (2003) *The Open Economy and the Environment: Development, Trade and Resources in Asia.* Edward Elgar, Chelthenham, UK, and Northampton, Massachussetts.

Coxhead, I., Rola, A. C. and Kwansoo, K. (2001) How do national markets and price policies affect land use at the forest margin? Evidence from the Philippines. *Land Economics* 77(2).

Coxhead, I., Shively, G. and Shuai, X. (2002) Development policies, resource constraints, and agricultural expansion on the Philippine land frontier. *Environment and Development Economics* 7, 341–363.

Cruz, M.C., Meyer, C.A., Repetto, R. and Woodward, R. (1992) *Population Growth, Poverty, and Environmental Stress: Case Studies from the Philippines and Costa Rica.* World Resources Institute, Washington, DC.

Cruz, W. and Repetto, R. (1992) *The Environmental Effects of Stabilization and Structural Adjustment Programs: The Philippines Case.* World Resources Institute, Washington, DC.

Deacon, R. (1995) Assessing the relationship between government policy and deforestation. *Journal of Environmental Economics and Management* 28, 1–18.

Deutsch, W.G., Busby, A.L., Orprecio, J., Cequina, E. and Bago, J. (1998) Community-based water quality indicators and public policy in the rural Philippines. In: Coxhead, I. and Buenavista, G. (eds)

Challenges of Natural Resource Management in a Rapidly Developing Economy: A Philippine Case Study. PCARRD, Los Baños, Laguna, Philippines.

Greene, W.H. (1993) *Econometric Analysis.* Prentice-Hall, Englewood Cliffs, New Jersey.

Jayasuriya, S. and Shand, R.T. (1986) Technical change and labor absorption in Asian agriculture: some emerging trends. *World Development* 14, 415–428.

Lopez, R. and Niklitschek, M. (1991) Dual economic growth in poor tropical areas. *Journal of Development Economics* 36, 189–211.

Mendoza M.S. and Rosegrant, M.W. (1995) Pricing behavior in Philippine corn markets: Implications for market efficiency. International Food Policy Research Institute Research Report 101. Washington, DC.

Midmore, D., Ramos, L., Hargrove, W., Poudel, D., Nissen, T., Agragan, F. and Betono, G. (1997) Stabilizing commercial vegetable production in the Manupali watershed, the Philippines. In: Coxhead, I. and Buenavista, G. (eds) *Challenges of Natural Resource Management in a Rapidly Developing Economy: A Philippine Case Study.* PCARRD, Los Baños, Laguna, Philippines.

NSO (Philippine National Statistics Office) (1990) *Bukidnon Provincial Profile.* NSO, Manila.

PCARRD (Philippine Council for Agricultural and Rural Resources Research and Development) (1999) *Guidelines for Watershed Management and Development in the Philippines.* PCARRD, Los Baños, Laguna.

Philippine Department of Agriculture (1994) *Grain Production Enhancement Program.* Philippine Department of Agriculture, Quezon City, Philippines.

Pingali, P. (1997) Agriculture-environment interactions in the Southeast Asian humid tropics. In: Vosti, S. and Reardon, T. (eds) *Sustainability, Growth and Poverty Alleviation: A Policy and Agroecological Perspective.* Johns Hopkins University Press for the International Food Policy Research Institute, Baltimore, Maryland.

Rola, A.C. and Coxhead, I. (1997) The agricultural economy of an upland community: a revisit to Lantapan, Bukidnon, Philippines. Manuscript, University of the Philippines at Los Baños, Philippines.

Shively, G.E. (1997) Consumption risk, farm characteristics, and soil conservation adoption among low-income farmers in the Philippines. *Agricultural Economics* 17, 165–177.

Shively, G.E. (1998) Economic policies and the environment: the case of tree planting on low-income farms in the Philippines. *Environment and Development Economics* 3, 83–104.

Silvapulle P. and Jayasuriya, S.K. (1994) Testing for Philippines rice market integration: a multiple cointegration approach. *Journal of Agricultural Economics* 45, 369–380.

Sims, C. (1980) Macroeconomics and reality. *Econometrica* 48, 1–48.

Southgate, D. (1988) The economics of land degradation in the third world. World Bank Environment Department, Working Paper No. 2. The World Bank, Washington, DC.

Tolentino, V.B.J. (1995) Intention vs. implementation of Philippine economic reforms under Aquino, 1986–1992. Institute of Developing Economies, VRF Series No. 240 (March), Tokyo.

5

How Do Relative Price Changes Alter Land Use Decisions?

Panel Data Evidence from the Manupali Watershed, Philippines*

I. COXHEAD[1] AND B. DEMEKE[2]

[1]*Department of Agricultural and Applied Economics, University of Wisconsin-Madison, 427 Lorch Street, Madison, WI 53706, USA, e-mail: coxhead@wisc.edu;* [2]*Department of Agricultural and Applied Economics, University of Wisconsin-Madison, 427 Lorch Street, Madison, WI 53706, USA, e-mail: demeke@aae.wisc.edu*

Introduction

This chapter asks the empirical question: 'How do farmers in remote, upper-watershed areas respond to price signals?' Until recently, it was widely assumed that most upland agriculture was undertaken primarily for subsistence. If correct, this would have important implications for the design of upland development programmes, as subsistence farmers, by definition, are beyond the reach of economic policies, and therefore, programmes addressing their needs would have to depend on direct interventions. Such interventions, including command and control policies to conserve natural resources, remain at the core of most resource conservation strategies in countries of the humid tropics.

In this respect, most upland development and conservation strategies lag behind the reality of areas at which they are directed. While pockets of pure subsistence production persist in least accessible regions, road and other infrastructure improvements have brought the majority of Asia's upland farmers into contact with markets and have thereby transformed the basis of their production decisions. Accumulating evidence indicates that farmers in remote areas are

*This chapter draws in part on Coxhead and Demeke (2004). We acknowledge with gratitude a long-standing collaboration with Dr Agnes Rola and her staff at the University of the Philippines at Los Baños, and the dedicated data collection efforts of Ms Isidra (Sid) Bagares, without whose efforts this research could not have been carried out.

increasingly willing to specialize in production of commercial crops and to rely on markets to supply household food needs. This engagement with the market increases the spatial and sectoral reach of economic policies, although obvious frictions in smallholder responses to policy changes may remain (e.g. Shively and Fisher, 2004). Market-based policies are generally cost-effective relative to direct interventions, so for policymakers, commercialization means that market instruments can be incorporated in the upland policy tool kit.

Continuing with the empirical focus of this book, this chapter explores land use decisions made by farmers in the Manupali River watershed. Our goal is to use data collected on upland farms to quantify responses to economic shocks, and thereby to understand the leverage that policies affecting agricultural prices can be expected to exert in upland areas. We use a decade-long panel of farm-level data on agricultural practices and prices. The existence of this data set makes possible a considerably more thorough examination and quantification of the market and policy determinants of land use decisions than is typically possible with cross-section data alone.

Markets, Policies and Land Use in the Uplands

As discussed in the previous chapters, in the Manupali River watershed the isolation of upland farmers from markets has diminished in the past generation, due in part to links created by migrants, and in part to infrastructural improvements. Using primary data from traders and wholesale markets, Coxhead *et al.* (see Chapter 4, this volume) show that farm markets in the watershed are integrated, and further, that local producer prices are determined exclusively by those in external markets. This finding forms one link in a chain relating farmers' actions to a set of macroeconomic determinants. The other link, which we now explore, is the one that connects farm-gate price signals to farmers' decisions. The empirical exploration of links between national markets and policies and upland land use decisions has only a few antecedents in the Philippines. Findings based on farm-level data have awaited the accumulation of time series, and are both very recent and somewhat tentative in their findings (Coxhead *et al.*, 2002; Shively and Pagiola, 2004).

In earlier work in the Manupali River watershed, Coxhead *et al.* (2002) examined farm-level land use decisions and related them to expected levels and variances of prices and yields, input prices, and quantities of fixed factors such as family labour. Their findings confirmed the price-responsiveness of land use decisions but suggested that price changes had relatively small land use effects. These results suggest a role for price policy, but not a highly influential one. Importantly however, the earlier study used a short series of observations (only three per farm); with highly heterogeneous farms, persistent differences between farms, rather than their common responses to market influences, dominated the results. In the earlier study, moreover, lack of wage data precluded formal labour market analysis. The analysis of this chapter extends that previous work in several important ways. We use a longer data series, and so can control for farm-specific effects. We now have a wage series and data on key crop price policies, and also take account of censoring and panel data properties in the data.

Over time, Philippine domestic prices of maize, vegetables and some other important upland crops have been subject to import-substitution policies applied as tariffs, state trading and/or outright import bans. These have raised producer prices substantially above world market equivalents (Coxhead, 1997; David, 2003). Under the terms of its accession to the WTO in 1995, the Philippines converted all such restrictions to a tariff equivalent form, but for some crops, the implied tariff has remained very high. This is the case for maize, the crop most widely grown in the watershed and in the Philippine uplands generally; the nominal protective rate (NPR) for maize has remained high, and even risen, since WTO accession in 1994 (Fig. 5.1).

Why is it important to quantify upland farmers' responses to trade policies? We hypothesize that these policies are responsible, at least in part, for the substitution of maize and some vegetable crops for other land uses, including perennials such as coffee, a shift that has increased deforestation and intensified the mining of upland soils. If so, then trade or price policy reforms that affect the national market for a crop could have widespread and potentially large aggregate environmental impacts achieved at low cost relative to the usual site-specific approach to sustainable upland development. Our goal is to discover whether prices, which transmit the effects of policy changes, induce significant land use responses once other determinants of behaviour are taken into account.

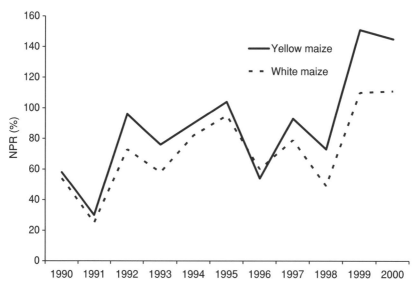

Fig. 5.1. Philippines: trends in the nominal protection rate for maize. (*Source*: ISPPS, University of the Philippines at Los Baños.)

Model

We assume that farm households choose land use strategies consistent with profit maximization over time, based on per-period net farm income. In the short run, they are constrained by the quantity of land they control, its characteristics (soil quality, climate, distance to market, etc.) and their labour endowment. In any period, the whole of the land area need not be farmed; some may be fallow. Household labour differs from hired labour on the farm in embodying specific management knowledge.[1]

By profit maximization, land in each period is fixed in total quantity but allocable across crops. Formally, define indexes i, j to correspond to farm products, k to inputs and h to observational units. Define land area per crop in period t by A_{it}, crop prices by p_{it}, wages and other input prices by w_{kt}, and a vector of variables whose values are specific to the household/farm or year by z_{ht}. Then we can write the area of each crop planted in each year as a function of these factors, namely $A_{iht} = A_{iht}(p_{jt}, w_{kt}, z_{ht})$. The choice of A_{iht}, for all crops i, is made by farmers who respond to economic incentives and are constrained by their endowments and sometimes by institutional features. Farm decisions are subject to exogenous changes in product and input prices and in exogenous or predetermined factors including farm size, family labour force and weather. At the intensive margin (i.e. within the current cropped area) farmers adjust labour and input use by crop. Between the intensive and extensive margins, they adjust land allocation among crops. At the extensive margin, they alter the total area of the farm that is cultivated in a given period.

Economically optimal behaviour would lead total farmed area to increase with output prices and decline with input prices, i.e. production costs. Allocation of land to individual crops should be positively related to expected own price or yield, and negatively related to prices or yields of other crops with which they compete for land and other resources. Higher wages are likely to reduce profitability and thus to reduce land demand; the extent of the wage effect should vary with the labour intensity of production. Increases in family labour are likely to increase land demand, especially in management-intensive crops, since experience indicates that family labour on the farm embodies specific managerial expertise.

In our data set, farmers allocate land mainly between maize and vegetable production. Vegetables are grown on small plots and are very input-dependent; hence, they are considerably more management-intensive than maize. Accordingly, whereas maize production can easily be expanded by hiring more labour (given land), the same may not be true of vegetables. Conversely, relaxing the land constraint (given family labour) should expand maize area, but may leave vegetable production unchanged if the household cannot provide a matching increase in managerial labour. These land and family labour constraints indicate a short-run model. Empirically, if these constraints bind, we can draw inferences about the incentives for farmers to take steps to relax them, following a shock of a given kind.[2] The same is true of farm credit, which is used almost exclusively by vegetable operations (see Coxhead *et al.*, 2002).

Data

Data come from a group of villages located in Lantapan municipality in the upper Manupali River watershed in central Bukidnon province, northern Mindanao Island. As discussed in Chapter 1, total area farmed in the municipality has increased faster than population, and this expansion has occurred at the expense of primary forest and other perennial crop and agroforestry systems. The main crop grown by area is maize; other crops vary in importance by location, with sugarcane dominant in the lower watershed and vegetables in the upper.

Researchers with the SANREM project conducted a series of farmer surveys in the dry seasons of even-numbered years from 1994 to 2002 – a total of five observations per household. Sample farms were drawn from nine villages in the middle and upper regions of the watershed (for survey methods and instruments, and summaries of findings from individual survey rounds, see Coxhead (1995) and Rola *et al.* (1995)). The surveys elicited information on household composition, farm production and input use, land use history, sales, non-farm income and farmers' perceptions of erosion and soil degradation, as well as their expectations on types of crop to be grown and crop prices in the next cropping season. The survey started with 190 farm households; in 1996 the sample size was reduced to 120, and another 20 households subsequently dropped out of the survey.

Land Use Responses to Economic Signals: Estimation and Results

Our goal in this section is to estimate land demand as a function of prices as well as farm and household characteristics. While several crops are grown in the study site, this analysis focuses on maize and vegetables, two enterprises for which land area per farm fluctuates from year to year. Table 5.1 provides a summary of the main variables used. In estimating land demand functions, the data present us with a number of practical challenges. First, the data are censored: many farms specialize (e.g. in maize production) and many observations thus record zero land allocation to particular crops in specific years. For maize, about 25% of total observations are zero; for vegetables, the figure is almost 70%. Hence, parameter estimates associated with land use changes that have been derived using standard linear regression techniques will be biased. Second, since the data form a panel, controlling for unobserved household-specific effects and dynamics is also essential. Third, price data are drawn from local market surveys and thus exhibit no cross-sectional variation across farms in each year. For a model containing several prices and few time periods, degrees of freedom become very limited. Fourth, land allocation decisions are arguably not independent, implying that single-equation estimation may not be efficient, since it does not take account of jointly made decisions on labour, other inputs and farm investments. Finally, as noted above, some households dropped out of the survey for various reasons, some of which are likely to be endogenous. When endogenous attrition of the sample is ignored, estimation captures only those farmers who remain in farming; hence the effect of prices and wages on abandoning farm lands could not be captured.

In this chapter, we adopt appropriate estimation methods to deal with parts of the first three issues (for details of our econometric approach, see Appendix, and Coxhead and Demeke (2004)). Since land allocation involves corner solutions, we use a Tobit estimator, which takes account of the necessity that total land use (expressed as a share of total area) be bounded between zero and one. However, there are additional complications. Empirical work using panel data must also deal with problems of unobserved heterogeneity across farms and through time, such as in soil quality, altered managerial ability and dynamic feedback effects. It has proved very difficult to address both dynamics and unobserved heterogeneity together (Hu, 2002); nor can we easily take account of the interdependence of land allocation equations. Hence, we restrict the analysis to a random effects Tobit model, leaving estimation of a system of dynamic equations for future work (see Demeke, 2004). The random effects Tobit model enables us to account for zero land allocation outcomes as well as for unobserved heterogeneity among the households.

We deal with the lack of cross-sectional variation in prices by substituting expected revenues in place of prices in the land demand functions. There are several studies that make use of expected revenue or expected return instead of output price as an explanatory variable (McGuirk and Mundlak, 1992; Holt, 1999; Arnberg, 2002). This approach merely requires the assumption that farmers respond equally to revenue-increasing shocks; whether the shocks originate in prices or yields is immaterial. Using production data, we estimate per hectare revenue per crop as a function of farm area, slope, education, distance, soil conservation, household labour, age, tenure, and location and

Table 5.1. Summary statistics of variables.

Variable	Mean	Std. Dev.
Total area planted (ha)[a]	1.16	1.45
Area of maize (ha)	1.02	1.46
Area of vegetable (ha)	0.14	0.33
Total farm area (ha)	2.68	2.95
Relative expected maize revenue (pesos/ha)	20.00	7.38
Relative expected vegetable revenue (pesos/ha)	44.80	30.08
Relative wage[b] (pesos/day)	0.17	0.02
Slope of the land (per cent)	16.38	10.70
Average distance from national road (km)	2.86	3.64
Years of education of the household head	6.28	3.43
Number of adults in the household	3.31	1.93
Age of the household head	44.39	11.79
Total number of observations	592	

[a] Sum of maize and vegetable area.
[b] Average wage in each village.

year dummies. We then predict expected revenues for all farms based on these estimates. This approach permits us to construct expected revenues for those farms with zero land allocation to a particular crop.

Assuming farm area to be exogenously fixed in the short run, each crop area allocation is postulated to depend on the following variables: total farm area, expected per hectare maize and vegetable revenues, wages, average slope of parcels, distance of the farm from national road, age and years of education of the household head, and number of adults in the family. In addition we estimate the effect of these variables on total area farmed. All price and value variables, that is, revenue and wages, are normalized by fertilizer price. Hence, the estimates capture the effects of relative price changes on land use decisions. Table 5.2 shows the random effects Tobit estimates of our model; Table 5.3 shows the key results as elasticities, which show the effect of a 1% increase in each exogenous variable on land area. These are computed at the sample means, and also at the means of land area observations that are greater than zero. From Table 5.3, crop area responses to expected revenues are positive for own-price and negative for cross-price terms. A 10% increase in maize price will induce farmers to increase maize area by 2.6%, reduce vegetable area by 12% and increase total planted area by 1%. Similarly, a 10% rise in the vegetable price will reduce maize area by 1.7%, increase vegetable area by 4.2% and reduce total area slightly, by 1.3%. Wage rises have negative effects on area planted, most especially so in the labour-intensive vegetable enterprise; a 10% wage rise reduces maize area by 2.3%, vegetable area by 23% and total area by 3.9%. Larger family labour

Table 5.2. Random effects Tobit estimates.

Variable	Maize		Vegetable		Total Area	
	Coef.	Std. Err.	Coef.	Std. Err.	Coef.	Std. Err.
Farm area	0.404***	0.020	−0.006	0.016	0.379***	0.014
Maize revenue	0.015**	0.008	−0.045***	0.009	0.006	0.007
Veg. revenue	−0.005**	0.002	0.007***	0.002	−0.004*	0.002
Wage	−1.570	1.939	−7.440***	1.814	−2.855*	1.634
Slope	−0.015**	0.006	−0.009*	0.005	−0.021***	0.005
Distance	−0.116***	0.018	−0.024	0.015	−0.107***	0.014
Educ. of HH	0.005	0.018	0.017	0.015	0.011	0.015
Family labour	−0.026	0.027	0.057***	0.023	0.002	0.023
Age of HH	−0.011*	0.006	−0.009*	0.005	−0.013***	0.005
Constant	1.032**	0.464	1.695***	0.404	1.689***	0.385
σ_c	1.037***	0.080	0.479***	0.063	0.935***	0.048
σ_u	0.810***	0.036	0.585***	0.038	0.703***	0.027
ρ	0.621	0.045	0.402	0.070	0.639	0.032
Log likelihood	−754.824		−340.862		−738.162	
Wald stat ($\chi^2(9)$)	491.06***		48.76**		909.01***	
N	580		580		580	

Note: *, **, and *** indicate significance at 1%, 5% and 10% levels respectively.

Table 5.3. Elasticity estimates from random effects Tobit model.

Variable	Maize area	Vegetable area	Total area
	Elasticity at sample means		
Farm area	0.905	−0.021	0.811
Maize revenue	0.258	−1.238	0.100
Veg. revenue	−0.169	0.422	−0.127
Wage	−0.225	−1.733	−0.391
Slope	−0.212	−0.200	−0.271
Distance	−0.276	−0.092	−0.245
Educ. of HH	0.029	0.150	0.054
Family labour	−0.071	0.257	0.007
Age of HH	−0.396	−0.538	−0.465
	Elasticity at means (if land allocation is positive)		
Farm area	0.665	−0.007	0.697
Maize revenue	0.189	−0.383	0.086
Veg. revenue	−0.124	0.131	−0.109
Wage	−0.165	−0.536	−0.336
Slope	−0.156	−0.062	−0.233
Distance	−0.203	−0.028	−0.211
Educ. of HH	0.021	0.046	0.046
Family labour	−0.053	0.079	0.006
Age of HH	−0.291	−0.166	−0.400

endowments are significantly associated with greater area planted for vegetables (where management is important) but not for maize. For the total area planted, increases in expected maize revenue cause area to increase, while wage rises reduce it.

These estimates, together with those for national to local market integration in Chapter 4, robustly complete the chain of causation between national markets or policies and upland land allocation decisions. In our data, output prices have a clear and unambiguous effect on demand for agricultural land: higher crop prices lead to area expansion, other things being equal, and to intensification of production on existing farms. Given that upper-watershed production is more land-intensive than in lower areas for the same crops, market or policy shocks that raise prices of maize or vegetables (i.e. that turn the terms of trade in favour of upland agriculture) result in reduced fallow and greater pressures for deforestation.

Higher wages also have a strongly negative effect on land demand. This result, together with that of family labour on vegetable area expansion, indicates that labour market events influence the expansion of upland farming in our study site. In the later years of the sample, strong non-farm job expansion outside the watershed has been associated with rising farm wages. This trend is now seen clearly to have slowed – and perhaps helped to reverse – the expansion and intensification of upland farming.

Discussion

Our econometric findings, which corroborate and substantially strengthen those in earlier research, have potentially important implications for understanding the effects of trade and agricultural policies, real wage growth and internal migration, and macroeconomic policies on land use at the cultivated frontier. These phenomena may all have direct and significant effects on the total area of upland agriculture, and also on the allocation of upland land by crop.

The results of this study are in turn important for the design of policies and projects directed at upland development and the conservation of forest, land and water resources. For example, WTO compliance requires that the Philippines scale back agricultural tariffs over time; the book value of the maize tariff is supposed to fall to 50% by 2005 from its current value of nearly 100%. By how much will this affect upland land use? If our findings are representative, tariff reduction in maize will reduce the domestic price by 25% and maize area by 6.5% – in our sample about 0.065 ha per farm, or about 45,000 ha in Mindanao and 225,000 ha nationally. Total upland farm area could fall by about 2.5%. If this land is retired into less erosive uses, these changes will substantially reduce deforestation pressures, loss of soil fertility and downstream or offsite damages. This effect will be even greater if the trends in world markets and/or Philippine trade policy raise demand for unskilled and semi-skilled labour in non-farm sectors. The Philippines is now an important supplier of labour-intensive goods such as garments, electrical machinery and components, and services such as call centres, to world markets. Expansion of these labour-intensive industries applies upward pressure to farm wages and encourages out-migration by members of farm families, both directly for employment, and indirectly for education and training. If wages rise by 25% – about half the rise seen over one decade in our data – maize area could fall by about 5.6% per farm, or almost 200,000 ha nationally, and total area farmed in uplands by almost 10%. This is a reduction in agricultural area comparable in magnitude to that experienced in Thailand during that country's economic boom from 1988 to 1996.

Conversely, the findings reported here also show that the success of efforts to work directly with farmers to change land use practices depends critically on the economic setting. If price signals indicate higher profits through expanded area and intensified cultivation, then direct interventions at the farm level – especially those involving additional costs, such as hedgerows or other conservation techniques, or substitution of lower-earning crops for higher – are much less likely to have an effect. Designers of watershed protection programmes must take the broader economic context into account.

Of course, some of the numbers we report have large standard errors. Even so, their magnitudes indicate clearly the potential for national-level price policy and market changes to alter demand for agricultural land in uplands. The commercialization of upland agriculture and the development of integrated national markets have thereby assisted in the creation of a powerful set of policy instruments by which to influence the use of scarce natural resources in the upland economy. Policy reforms, and economic growth in general, are likely to have much greater, longer-lasting and more cost-effective impacts on the management of the upland resource base than are many location-specific 'sustainable agriculture' projects, replicated many times over. Working with policy designers as well as with farmers and local communities is an essential mix if watershed protection programmes are to have the best chances of success.

Notes

[1] While we observe this in our data, farmers give many reasons for fallowing, sometimes citing soil fertility and at other times the difficulty of obtaining seed, inputs or labour. Our model does not attempt directly to model the fallowing decision, nor soil quality dynamics; rather, we restrict ourselves to asking what governs the choice of land cultivated to crops.
[2] In the later years of our panel, introduction of banana plantations encouraged some households to let land. Extending the model to include this rental decision is a refinement that we address in ongoing research.

References

Arnberg, S. (2002) *Estimating Land Allocation Using Micro Panel Data Controlling for Yield Expectations and Crop Rotation*. SOEM-publication no. 47. AKF Forlaget, Copenhagen.

Coxhead, I. (1995) A baseline economic survey of the Lantapan agricultural economy. SANREM CRSP Philippines: Social Science Working Paper No. 95-1.

Coxhead, I. (1997) Induced innovation and land degradation in developing countries. *Australian Journal of Agricultural and Resource Economics* 41, 305–332.

Coxhead, I. and Demeke, B. (2004) Panel data evidence on upland agricultural land use in the Philippines: can economic policy reforms reduce environmental damages? *American Journal of Agricultural Economics*, 86, 1354–1360.

Coxhead, I., Rola, A.C. and Kim, K. (2001) How do national markets and price policies affect land use at the forest margin? Evidence from the Philippines. *Land Economics* 77, 250–267.

Coxhead, I., Shively, G.E. and Shuai, X. (2002) Development policies, resource constraints, and agricultural expansion on the Philippine land frontier. *Environment and Development Economics* 7, 341–363.

David, C.C. (2003) Agriculture. In: Balisacan, A. and Hill, H. (eds) *The Philippine Economy: Development, Policies, and Challenges*. Oxford University Press, New York, and Ateneo de Manila University Press, Quezon City, Philippines, pp. 175–218.

Demeke, B. (2004) How do national policies and local labor-market changes affect land use decisions in the fragile uplands of the Philippines? An application of Amemiya-Tobin multivariate censored systems of equations with panel data. Manuscript, University of Wisconsin-Madison, Wisconsin.

Holt, M.T. (1999) A linear approximate acreage allocation model. *Journal of Agricultural and Resource Economics* 24, 383–397.

Hu, L. (2002) Estimation of a censored dynamic panel data model. *Econometrica* 70, 2499–2517.

McGuirk, A. and Mundlak, Y. (1992) The transformation of Punjab agriculture: a choice of technique approach. *American Journal of Agricultural Economics* 74, 132–143.

Rola, A.C., Elazegui, D.D., Paunlagui, M.M. and Tagarino, E.P. 1995 Socio-demographic, economic, and technological profile of farm households in Lantapan, Bukidnon, Philippines: 1995. SANREM CRSP-Philippines Research Report, July 1995.

Shively, G.E. and Fisher, M.M. (2004) Smallholder labor and deforestation: a systems approach. *American Journal of Agricultural Economics* 86, 1361–1366.

Shively, G.E. and Pagiola, S. (2004) Agricultural intensification, local labor markets, and deforestation in the Philippines. *Environment and Development Economics* 9, 241–266.

Shonkwiler, J.S. and Yen, S.T. (1999) Two-step estimation of a censored system of equations. *American Journal of Agricultural Economics* 81, 972–982.

Wooldridge, J. (2002) *Econometric Analysis of Cross Section and Panel Data.* MIT Press, Cambridge, Massachusetts.

Appendix

This appendix describes the econometric method used in the analysis. Assuming strict exogeneity, for each crop we estimate the following model (suppressing crop subscripts):

$$A_{it} = \max(0, x_{it}\beta + c_i + u_{it})$$
$$u_{it} \mid x_i, c_i \sim Normal(0, \sigma_u^2)$$
$$c_i \mid x_i \sim Normal(0, \sigma_c^2)$$

for $I = 1, \ldots, N$ households, and $t = 1, \ldots, T$, periods, where A_{it} is land area planted to maize or vegetables, x_{it} are explanatory variables and β, σ_u^2, and σ_c^2 are parameters to be estimated. We also estimate an equivalent model for total crop area. Each dependent variable is estimated as a function of prices, wages, technology and other conditioning variables.[1] We assume that total farm area is exogenously fixed in the short run. Each land allocation equation is assumed independent, and we estimate the model equation by equation.[2]

Estimation of marginal effects in the random effects Tobit model poses its own problems, since these by definition involve unobserved individual effects. In practice, marginal effects are estimated by making normalization of the individual-specific effects such as $E(c) = 0$ (Wooldridge, 2002: 22). It follows that:

$$\frac{\partial E(A_{it} \mid x = \bar{x}, c = 0)}{\partial x_{jt}} = \Phi\left(\frac{\hat{\beta}\bar{x}_{it}}{\sigma_u}\right)\beta_{ij},$$

where $\bar{x}_{jt}$ is the sample mean and Φ is the CDF of a standard normal distribution.

Notes

[1] We estimate the model using the *xttobit* procedure in STATA version 7.
[2] The independence assumption leads to an efficiency loss. One can use the two-step method of Shonkwiler and Yen (1999) to overcome this drawback; however, this approach does not account for interdependence between land use choices in the first stage. A better approach would be to estimate the systems of equation using Maximum Simulated Likelihood method (see Demeke, 2004, for this approach).

6   Economic Incentives and Agricultural Outcomes in Upland Settings*

I. COXHEAD[1] AND G. SHIVELY[2]

[1]Department of Agricultural and Applied Economics, University of Wisconsin-Madison, 427 Lorch Street, Madison, WI 53706, USA; e-mail: coxhead@wisc.edu; [2]Department of Agricultural Economics, Purdue University, 403 West State Street, West Lafayette, IN 47907, USA; e-mail: shivelyg@purdue.edu

Agricultural Intensification and Land Degradation

Changes in agricultural land use patterns in tropical watersheds such as the Manupali River watershed are clearly related to economy-wide phenomena, including relative prices and market conditions (Coxhead and Jayasuriya, 1995; Shively, 1998). Generally speaking, many of the most important incentives that shape land use in watersheds are determined well beyond the borders of the watershed. To study the connections between key economic variables and changes in land and resource allocation at the farm level, economists typically employ either partial equilibrium or general equilibrium models. Partial equilibrium approaches investigate changes in the agricultural sector in isolation from other sectors of the economy. A partial equilibrium approach can be very useful, but can sometimes produce misleading results, especially if the agricultural sector is relatively large in relation to labour markets, private consumption expenditures and international trade, as it is in many developing countries (Coxhead and Warr 1995; Coxhead, 1997). In these settings, changes that occur in agriculture can easily affect other sectors of the economy, and vice versa, by inducing changes in the prices of factors (inputs) and commodities (outputs) used in other sectors. General equilibrium models have the advantage that they can account for these economy-wide linkages when used to study a specific part of the economy, such as the agricultural sector.

*Portions of this chapter appeared previously as Coxhead, Ian A. and Shively, Gerald E. (1998) Some Economic and Environmental Implications of Technical Progress in Philippine Corn Agriculture: an Economy-wide Perspective, *Journal of Agricultural Economics and Development* 26, 60–90. This material appears with the permission of that journal.

In this chapter we present a framework for measuring the changes in economic and environmental outcomes that occur in watersheds subject to land degradation, both in partial and in general equilibrium. We use this framework to calculate changes in upland land degradation due to changes in economic conditions.[1] We present and discuss simulation resul. for an applied general equilibrium (AGE) experiment that examines the likely economic and soil erosion implications of technical progress in maize production in the Philippines. We focus on maize, due to its widespread appearance in the upper reaches of many tropical watersheds, not just in the Philippines. We estimate the value of changes in soil degradation resulting from exogenous changes occurring both within upland agriculture and in other sectors of the economy. To do this we combine nutrient replacement cost estimates with data on annual soil erosion rates for different upland crops in the Philippines. We use this information, together with data on upland crop area, to calculate the aggregate value of changes in land degradation due to changes in production technologies and land area planted to upland crops.[2]

Measuring the Impact of Land Degradation in General Equilibrium

In standard general equilibrium models, growth rates of effective factor endowments are typically taken to be exogenous and each factor is regarded as being of uniform quality. However, for two reasons these assumptions may not be supported when agricultural intensification occurs in the upper reaches of tropical watersheds. First, without compensating investments, more intensive cultivation or an expansion of cultivated area in upland areas is usually associated with a decline in average soil quality. Second, different land uses may be associated with different rates of land degradation on existing agricultural land. As a result, an attempt to assess the economic value of changes in agricultural land use should incorporate on-site land degradation costs. In the Appendix to this chapter we describe how the structure of a general equilibrium model can be modified to allow land quality to reflect agricultural land use. We use our model to explore the welfare and environmental implications of some changes in agricultural prices and technology via a series of simulation experiments.

Our approach takes account of the fact that, in addition to resource stocks, key commodity prices themselves may be determined at least in part by supply-and-demand shifts in the domestic economy.[3] Endogenous price formation is an important consideration when the agricultural sector is large in relation to factor and product markets, trade, household expenditures or government revenues, as in the Philippines. Moreover, the same arguments for a general equilibrium approach apply in factor markets. For example, the marginal value product of agricultural land depends not only on output prices and technology, but also on prices of substitutable inputs, such as labour. When labour is mobile across sectors, a rise in non-agricultural wages promotes reallocation of land to relatively less labour-intensive crops. By implication, changes that alter relative factor prices within a watershed could encourage the reallocation of land towards or away from relatively erosive crops.

Our simulation analysis is based on a modification of the Agricultural Policy Experiments (APEX) model of the Philippine economy. APEX is an applied general equilibrium model of the Philippine economy developed in a collaborative venture by researchers at the Australian National University and the Philippine Department of Agriculture. APEX is a conventional, real, micro-theoretic general equilibrium model designed to address microeconomic policy issues for the Philippines. All parameters describing technology and preferences have been constructed from econometric estimates.[4] The model contains 50 producer goods and services produced in 41 industries. These are aggregated into seven consumer goods. The model consists of five households, each representing a quintile of the income distribution and having unique income and consumption characteristics based on Philippine household income and expenditure data. Producer demands for factors (land, labour and capital), aggregation over factors of different types, and consumer demands for goods and services are all described by flexible functional forms. Imports and their domestically produced substitutes are aggregated using CES forms with econometrically estimated Armington elasticities.[5]

Agricultural production in the model takes place in three regions of the Philippines (Luzon, Visayas and Mindanao). The agricultural sector produces a set of products using land, unskilled labour and fertilizer inputs. Land is specific to agricultural uses, while the other inputs are not. Some groups of agricultural goods are presumed to be jointly produced. One such group is the category 'rainfed crops', which includes rainfed rice, maize and root crops. This sub-aggregate is the set of agricultural crops that are ubiquitous in the upper reaches of most Philippine watersheds, and the ones in which there is potential for measurable soil quality depletion through erosion. According to the APEX database, value-added in the rainfed crops sector is dominated by maize (60%), with root crops accounting for 28% and rainfed rice 12%.

In our model, the joint production function for rainfed crops is nested within that for agriculture as a whole in each region. From the perspective of understanding the influence of economic forces on land use changes in a watershed, the key feature of the model is that the composition of production within the rainfed crops sector (and by extension, on vulnerable upland soils) is influenced by the relative prices of the three crops, or by crop-specific technical progress. Similarly, the share of rainfed crops in total agricultural production (and therefore the importance of upland production relative to other forms of agriculture) depends on prices and rates of technical progress of the sub-aggregate relative to those of other agricultural sectors.

To study land degradation using the APEX model we require estimates of the crop and region erosion weights. All other data are drawn from the APEX database. Table 6.1 contains our estimates of land use in Philippine uplands. Using these estimates, together with data on annual soil loss per hectare associated with these uses (Table 6.2), we compute total soil loss for each category of upland land use at two slope classifications (Table 6.3). We allocate these across regions using regional shares of agricultural land use by crop. Final values of soil loss and regional weights, calculated from the data in Tables 6.1–6.3, are presented in Table 6.4. Estimates of erosion rates in maize and upland rice take account of crop–fallow systems in uplands characterized by very high erosion rates during periods of cultivation, followed by declining erosion rates as the length of fallow increases (for details see Coxhead and Shively, 1995). In the model we use estimates only for upland rice and upland maize, ignoring erosion changes associated with land of less than 18% slope, and other forms of agricultural and non-agricultural activity which can not be reliably linked to output changes by sector.

Table 6.1. Land use in the Philippine uplands (ha).

| Land use | Slope category (%) | | Total |
	18–30	30+	
Rice	315,000	52,500	367,500
Maize	375,000	61,250	436,250
Fallow	3,970,000	1,540,000	5,510,000
Other agriculture	592,000	96,250	688,250
Non-agricultural (forest)	–	–	7,900,000
All uses	–	–	14,902,000

Source: Based on data in World Bank (1989), Annexes 1–2.

Table 6.2. Soil loss for various land uses and slopes (tonnes/ha/year).

| Land use | Slope category (%) | |
	18–30	30+
Rice	50	100
Maize with fallow	50	150
Other agriculture	25	50
Forest	1	1

Source: Authors' estimates from secondary data. For details see Coxhead and Shively (1995).

Table 6.3. Estimated total soil losses for land uses and slopes (tonnes/year).

| Land use | Slope category (%) | | Total |
	18–30	30+	
Rice	15,750,000	5,250,000	21,000,000
Maize with fallow	217,250,000	240,190,000	457,340,000
Other agriculture	14,800,000	4,812,500	19,612,500
Non-agricultural (forest)	–	–	7,900,000
All uses	–	–	505,952,500

Source: Computed from data in Tables 6.1 and 6.2.

Table 6.4. Weights used in regional estimates of erosion shares, by sector.

Sector shares in total erosion (α)	Region		
	Luzon	Visayas	Mindanao
Upland rice	0.09	0.04	0.02
Maize	0.83	0.88	0.95
Other agriculture	0.06	0.06	0.02
Non-agricultural (forest)	0.02	0.02	0.01
Regional shares in national erosion (λ)	232	0.234	0.534

Source: Computed from data in Table 6.3. Regional land use shares derived from the 1980 Census of Agriculture (NSO, 1981).

In the simulation analysis we allow changes in land allocation to different crops (and in some cases total land area) to respond to changes in commodity and factor prices, but maintain the restriction that rates of technical progress and land management practices are exogenous. Even with these restrictions, the environmental, distributional and welfare implications of a given exogenous change cannot easily be predicted *a priori*, because prices, incomes and land allocation are all simultaneously determined. This suggests that policymakers focusing on sector-specific solutions to land degradation problems in upper watersheds may be overlooking some important indirect sources of change in agricultural resource allocation, originating in non-agricultural sectors. Conversely, policy interventions addressing non-environmental goals may have environmental implications in critical watersheds. We explore both of these points in the simulation experiments and in our discussion of the results.

Simulation Results

In this section we use simulation results obtained from the model to assess the environmental consequences (as captured by changes in on-site agricultural land degradation) of technical progress in maize. The policy setting of these experiments is the long-standing Philippine government goal of promoting domestic staple grain production by subsidizing R&D and extension efforts aimed at increasing productivity in rice and maize production (David, 2003). Given the resources devoted to this effort and its potential for inducing land use change, we use the APEX model to simulate maize yield changes in the 1990s. This permits us to decompose their outcomes and to quantify environmental as well as welfare impacts.

The policy setting

Alternative closure assumptions regarding the macroeconomic setting in which farmers operate are possible in the APEX model. Due to uncertainty about the true nature of the economy, we employ three alternatives. All share some characteristics: international prices of imports and exports are exogenously fixed (i.e. we assume the Philippines is a small country in relation to the international economy); the nominal exchange rate is also fixed, providing a *numeraire* for domestic prices. The current account, budget deficit and real savings of households are also fixed, so the effects of exogenous shocks are fully absorbed by current-period changes in real household expenditures. The government budget is balanced by endogenous adjustments in the rate of a lump sum tax on households. What distinguishes our three alternative views of the macroeconomy is the way in which we characterize two markets: (i) for land; and (ii) for grains and grain products.

In the first closure the quantity of agricultural land is fixed and its price is determined by changes in agricultural profitability. Grain prices are endogenous, trade in rice and maize is unrestricted (although the government levies *ad valorem* tariffs on imports), and government grain demand is exogenously fixed. This is a standard neoclassical closure. Given its characteristics, we denote this as the *unrestricted* case.

In the second and third closures we impose different sets of rules on grain markets and on the regional land markets. Erosion in Philippine agriculture is largely associated with production of rice and maize in rainfed and upland areas, and the government has historically exerted a broad influence on grain prices and demand. Our choices are motivated by the need to capture the effects of these interventions. Before describing the second and third closures, we present some historical information on grain market policies.

Since the 1950s international trade, retail prices and some storage and processing of rice, maize and wheat in the Philippines have been administered by the body now known as the National Food Authority (NFA). The NFA's legislated objective is to maintain price and supply stability in grains, although its interventions have always been understood to serve an additional goal, that of promoting distributional equity. The NFA strives to support prices paid to rice and maize farmers whilst reducing food costs for urban consumers (Intal and Power, 1990). The NFA's instruments have included a monopoly on international trade, nominal consumer price controls and fixed capital investments, including construction and operation of facilities for the purchase, milling and storage of rice and maize. Imports and changes in the NFA's domestic stocks have been used to defend legislated producer price floors and consumer price ceilings. The NFA's activities are not self-financing; the agency is supported by Philippine government subsidies.[6]

Given these subsidies, the NFA has not been entirely successful in defending consumer price targets and producer prices have consistently fallen below the NFA floor during periods of rapid agricultural productivity growth, in spite of massive NFA stockpiling (Lantican and Unnevehr, 1987; Shively *et al.*, 2002).[7] Another side effect of domestic interventions has been low private profitability in the grain milling and storage industry; the NFA has used its subsidy to finance a low or negative net margin between producer and consumer prices (Lantican and Unnevehr, 1987; Shively *et al.*, 2002). This strategy has discouraged private entry and investment in grain milling, and at times has led the government to offer countervailing subsidies to help restore profitability and promote investment in the sector.

To capture these features within APEX we adopt the following modifications to the unrestricted closure described above. We fix rice and maize imports exogenously,[8] and make government purchases of the output of the rice and maize milling sector endogenous. With these changes there is no endogenous international trade in rice and maize, so when domestic supply increases more rapidly than demand the excess is bought by the government at a fixed price and stockpiled. As a result nominal consumer prices of cereals are also fixed. Producer prices of rough rice and maize are strongly influenced (although not exclusively determined) by demand from the rice and maize milling sector from which the government makes its grain purchases.[9]

The second and third closures both employ this representation of grain markets, contrasting with the unrestricted closure described above. They differ from one another in that the second case (*NFA-fixed land*) holds total agricultural land area fixed, while the third case (*NFA-fallow*) makes land supply in each region endogenous but holds price fixed. In this closure, land may be brought into or removed from production in response to changes in demand. Table 6.5 summarizes the key features of the three closures.

Technical progress in rainfed crops

In the *Philippine Agricultural Development Plan (1991–95)* (Philippine Department of Agriculture, 1990), domestic maize production was targeted to grow by 5.71% per year from 1991 to 1995. Support for this growth was to be provided through a combination of price supports and technical assistance to maize farmers, especially in provinces designated as Key Production Areas (KPAs). The projected growth rate was consistent with the 1950–1985 average of 5.8% per year (Intal and Power, 1990). Historically, however, more than two-thirds of the growth in domestic maize production came from increases in

Table 6.5. Key features of three alternative closures.

Variable	Unrestricted	NFA-fixed land	NFA-fallow
Land Market			
Agricultural land area	Exogenous	Exogenous	Endogenous
Return to land	Endogenous	Endogenous	Exogenous
Grain Markets			
Rice and maize imports	Endogenous	Exogenous	Exogenous
Rice and maize trade shifter	Exogenous	Endogenous	Endogenous
Government purchases[a]	Exogenous	Endogenous	Endogenous
User prices[a]	Endogenous	Exogenous	Exogenous

[a] In the rice and maize milling (RCM) sector.

maize area, and only about one third came from yield increases (Intal and Power, 1990). From a watershed perspective, area expansion typically takes place in upper reaches of the agricultural landscape, at the so-called extensive margin. Thus, even if half the targeted output growth were realized through higher yields (implying an historically high rate of technical progress for Philippine maize), the plan would still have required the area planted to expand by about 16% over the 5-year period, and most of this expansion would have taken place on steeply sloping land. Maize production is widely regarded as one of the most erosive agricultural uses of sloping land in the tropics and among upland crops in the Philippines, the chief contributor to soil nutrient depletion (David, 1988). This rate of maize area expansion therefore would imply a significant increase in total agricultural land degradation.

What were the likely economic and environmental effects of technical progress in maize production? In our experiment we simulate a 10% rate of factor-neutral technical progress in this sector. Initially we assume that resulting economic and environmental changes in maize production, demand and price would have been determined independently of government interventions and adopt the *unrestricted* closure. We then compare the results with those from the alternative characterizations of grain and land markets in the *NFA-fixed land* and *NFA-fallow* closures.

Key economic results of this experiment are presented in Coxhead and Shively (1998), where we show that a 10% increase in the rate of technical progress in maize leads to a less than proportional increase in output (1.9%). Grain imports also decline: rice by 10.5% and maize by 3.7%. These are the expected outcomes given that the average household expenditure elasticity of demand for cereals is less than 0.5 (Balisacan, 1992), since final demands dominate intermediate use of maize for feed by livestock sectors. The productivity gain does not lead to increased sales, but rather to a substantial drop in price (23%). By driving down prices, technical change thus causes resources to be drawn out of the maize sector: Table 6.6 shows imputed regional land demand for maize production falling by 3–6%.

Since erosion rates in maize are higher than in other crops, the extent of the price decline creates a somewhat surprising environmental result: land degradation rates drop substantially, by an average of 4.3% over all regions. By region, the decline is greatest (5.5%) in Mindanao, where maize occupies about 41% of agricultural land, and smallest (2.8%) in Luzon, where it occupies 18%. In the Visayas, were maize is initially grown on 32% of agricultural land; erosion declines by an intermediate amount (3%). Technical progress in maize, by driving down producer prices, has caused some agricultural land to be shifted from maize production to less erosive uses and therefore has had a beneficial impact on agricultural land degradation.

When the cost of land degradation is ignored, technical progress in maize raises real aggregate household consumption by 1.5% (Coxhead and Shively, 1998). However, since technical progress in maize also reduces the rate of erosion, it follows that technical progress increases aggregate welfare if degradation costs are counted. However, the change has a regressive effect on income distribution, and this is important information for policy purposes. In upper-

Table 6.6. Technical progress in maize: effects on agricultural land use and erosion.

		Closure	
	Unrestricted	NFA with fixed land	NFA with fallow land
Regional land use changes			
Luzon			
Irrigated palay	4.09	5.57	2.92
Non-irrigated palay	0.34	1.51	−1.01
Maize	−2.98	−2.50	−4.78
Coconut	4.88	2.49	0.17
Sugarcane	4.62	2.37	0.04
Visayas			
Irrigated palay	3.61	6.04	5.94
Non-irrigated palay	1.27	2.63	2.63
Maize	−3.47	−2.94	−2.59
Coconut	5.00	4.28	4.52
Sugarcane	4.40	3.56	3.78
Mindanao			
Irrigated palay	9.22	11.22	6.96
Non-irrigated palay	−0.47	1.35	−2.65
Maize	−6.02	−5.06	−8.65
Coconut	6.49	3.90	0.07
Sugarcane	10.61	8.90	4.95
Regional fallow area changes			
Luzon	0.00*	0.00*	1.88
Visayas	0.00*	0.00*	−0.19
Mindanao	0.00*	0.00*	3.14
Erosion changes			
Luzon	−2.75	−2.34	−4.46
Visayas	−2.95	−2.51	−2.19
Mindanao	−5.55	−4.60	−8.13
Total	−4.29	−3.59	−5.89

* indicates the level is fixed exogenously.

watershed areas, where the poorest of the rural poor are found, soils are generally poor, market access relatively weak, credit markets are largely absent and information flows sporadic. For all these reasons, farmers in these more remote areas are less likely to be able to take advantage of technological innovations. The price effects due to the economy-wide increase in maize supply would, however, reach these farmers, as the analysis in Chapter 4 has established. Therefore (although we cannot read this directly from the model), upper-watershed farmers are likely to experience little of the gain from technical progress, but all of the consequent price decline, and their profits and household incomes will fall. Though it is possible that part of the loss would be compensated by job creation in more favoured areas, the net effect is still most likely to be negative, and could in turn trigger a land use response that exacerbates environmental pressures.

The results in the unrestricted closure assume that the drop in the maize price generated by technical progress does not trigger action to support producer prices. Price supports could diminish and perhaps reverse our maize production and land degradation results. We examine this possibility by repeating the technical change experiment using a closure in which the government (through the NFA) absorbs excess supply of cereals by buying the output of the grain-milling sector at a fixed price.[10] As expected, more elastic demand for grains dampens the price-reducing effect of the technical change: the producer price decline (−19.5%) is only 85% of that observed in the unrestricted closure. Maize output thus rises twice as fast as in the unrestricted closure (3.7% versus 1.9%); less land is drawn out of maize production, and erosion declines by smaller amounts in each region. The national average erosion decline, −3.6%, is only 84% of that observed in the first experiment.

The additional resources attracted into rice and maize production contribute to modest declines or reduced growth rates in the output of services, natural resources and manufacturing sectors, and boost agricultural processing industries (Coxhead and Shively, 1998). However, it is on the consumer side of the economy that the NFA intervention is most strongly felt. With price supports in place, consumer prices of cereal products decline by 1.5% relative to the CPI. The relative distribution of gains and losses among households remains similar, but the aggregate welfare gain from the technical progress, 0.9%, is only two-thirds of that indicated in the unrestricted case.[11]

Part of the reallocation of land arises as a result of the assumption maintained in the unrestricted and NFA-fixed land closures that total land area remains fixed. Because total land area is fixed, an expansion or contraction of maize land due to technical progress must be matched by a corresponding contraction or expansion of land used for other crops. But empirically, it is much more likely that some of the land used for maize production, especially in upper reaches of watersheds, is of such low quality that farmers reducing their maize acreage would allow the land to revert to fallow rather than plant new crops.[12]

Our results change when we fix the nominal return to agricultural land and allow the area to adjust endogenously. In Table 6.6 we see that some of the land removed from maize production is fallowed (about 2% averaged over all regions), and that land use and output growth in other agricultural sectors are correspondingly lower. As a result of the removal of some land from production, aggregate erosion declines much more rapidly than in the fixed land case. Other results are broadly comparable between the two NFA closures, except that when land area is endogenous, measured real GDP and welfare increase by less; some resources are taken out of production rather than being reallocated to other uses within the agricultural economy.

The two NFA closures provide rough upper and lower bounds for our estimates of the predicted change in erosion. The NFA-fixed land closure, by requiring that all land taken out of production of one crop be immediately planted to another, overstates the capacity of the agriculture sector to adjust land use in response to a relative price change. The NFA-fallow closure assumes land can be brought into production at zero cost, and thus understates the constraints to expansion at the extensive margin. Approximately 7% of arable land in the Philippines is fallow (NSO, 1981); therefore, of the two NFA closures the NFA-fallow variant seems more likely to be representative of actual Philippine conditions.

In our model, the inelasticity of domestic demand for cereals creates a treadmill effect, in which productivity gains won by maize producers are more than offset by terms of trade losses. As long as land can be reallocated to less erosive uses (including fallow), this inelasticity has a positive environmental effect, since falling prices cause maize area to decline. This decline occurs whether or not there is intervention in the grains market. However, to the extent that interventions dampen endogenous price effects, they reduce the probability of significant environmental gains from technical progress in maize. Moreover, cereal market interventions do little to improve the distribution of real household expenditures, and impose efficiency costs on the entire economy.

Discussion

Valuing changes in soil erosion

We now locate our simulation results in a wider economic and welfare context by applying monetary values to the land degradation changes predicted by the model. Full valuation of soil erosion is extremely difficult because many off-site damages evolve slowly over time. In this study we maintain a conservative stance by accounting for on-site losses only, and by considering changes in erosion over base levels for two upland activities only: rainfed rice and maize. Combined, these sectors account for roughly 42% of land area in Philippine uplands, but about 90% of upland agricultural soil loss (Table 6.4).

We calculate the aggregate value of changes in land degradation (D) by the formula $D = e \cdot E \cdot P$, where P is the annual estimated value of erosion (per tonne), and E and e are initial aggregate erosion and its proportional change as defined in equation (6.10) of the appendix. Since erosion estimates provided by the simulations are expressed as percentage changes over a base level, we combine the base estimates of soil loss (the central figures in Table 6.3) with the model's estimates of percentage changes in erosion to calculate total soil losses. Combining figures for rice and maize lands at all slopes over 18% provides an estimate for E of 478 million t of soil loss per year. For valuations we rely on Cruz et al. (1988). Using experimental plot data from the Magat and Pantabagan watersheds of Luzon, these authors computed losses from erosion for three critical soil nutrients.[13] We adopt their cost estimate of nutrient loss of 15.46 pesos ($0.60) per tonne of eroded soil. Substituting this value into equation (6.11) yields the aggregate costs and savings reported in the initial rows of Table 6.7.

The simulation experiment produced a reduction in land degradation, as land formerly used to grow erosive cereal crops was switched to other uses, including fallow. The reduction in land degradation produces gains (or rather, reduced losses) to the economy of US$10–17 million, depending on which

Table 6.7. Value of reduced land degradation due to technical progress in maize.

	Closure		
	Unrestricted	NFA-fixed	NFA-fallow
Percentage change in erosion	-4.29	-3.59	-5.89
Total change in erosion (million tonnes)	-20.50	-17.20	-28.20
Total savings (million US$)	-12.30	-10.30	-16.90
Savings as % of 1991 GDP	0.04	0.04	0.06
Savings as % of 1991 GDP growth	3.18	2.66	4.36
Savings as % of agricultural value added	0.20	0.17	0.27
Savings as % of govt. expenditures on agric.	1.58	1.33	2.17
Savings as % of the environmental component of govt. expenditures on agric.[a]	21.12	17.68	29.02

Notes: Base data from *Philippine Statistical Yearbook 1993* except [a]–1.4 billion pesos at 1980 prices (David *et al.*, 1992).

closure is adopted. While amounts are small in relation to total GDP, they represent approximately 0.3% of value added in agriculture. They are equal to approximately 4% of government expenditures on agriculture and about one-fifth of the 'environment' component of the Department of Agriculture budget. The reductions in land degradation losses are large relative to the GDP growth gains predicted to occur as a result of technical progress.[14]

Conclusion

This chapter has highlighted the importance of general equilibrium economic linkages for assessments of the environmental impacts of economic changes. In terms of the overall costs of upland soil erosion in the Philippines, our approach provides very conservative estimates. We did not account for off-site effects, and in constructing our database we consistently used 'low' estimates when parameter values were uncertain. The analysis nevertheless provides insights into the ways in which watershed integrity is affected by changes in the general economy, especially in places where cultivation decisions are being made at the extensive margin.

The results of this work indicate that not all forms of agricultural growth necessarily increase land degradation. In the Philippines, investment in technical progress in maize – a relatively erosive crop characterized by inelastic demand – results in reduced maize area and thus less land degradation, so long as the gains from productivity growth result in lower producer prices. Productivity growth has a regressive impact on real household expenditures, but only in the sense that the real expenditures of the poor rise less rapidly than those of the rich. In such circumstances government interventions to support producer prices are likely to have only modest success in maintaining producer incomes, to have little ameliorative impact on income distribution, and may also diminish the environmental benefits of the productivity gain. Indeed, the analogy between technical progress and a price rise underlines the potential environmental costs of price supports for producers of land-degrading crops, a point explored more fully in Coxhead and Jayasuriya (2004).

Watershed-level environmental gains from productivity growth also depend on land in upper-watershed areas being removed from maize production. The conditions under which this might take place vary across locations. In particular, gains by the poor in relatively remote areas are contingent on their mobility in the labour market. Mobility in turn depends on skills, information flows and migration costs; security of tenure may also be influential when land – an important form of household wealth in poor economies – cannot be retained with certainty when household members leave to work for extended periods elsewhere.

In much of Asia, the 1990s saw substantial increases in job creation not only in major cities but also, to an increasing extent, in peri-urban and rural areas. Rural industrialization and the growth of regional cities in South-east Asia, together with better and cheaper transport links, have reduced migration costs; for increasing numbers of farm households, it is now feasible to commute to non-agricultural jobs without changing residence. Thus intersectoral migration possibilities have increased, providing alternative incomes at least for some upland farm households (Rola and Coxhead, 2003). Conversely, when mobility is low, other mechanisms are needed if upper-watershed farmers are to change land use and/or be compensated for price declines that would otherwise increase their poverty (see Pagiola *et al.*, Chapter 11, this volume).

Notes

[1] Land degradation (used interchangeably with soil erosion in this chapter) denotes on-site fertility losses due to nutrient loss and damage to soil structure. While exact quantification is extremely difficult, the World Bank (1989) estimated the value of on-site fertility losses – land degradation – attributable to upland agriculture in the Philippines at US$100 million per year.
[2] While it refers to a real phenomenon in the Philippine context, the focus on technical progress as opposed to other sources of change is adopted for convenience. The effects of this change are broadly similar to those of a producer price rise. For analyses of the effects of price changes, including those induced by policy reforms, see Strutt and Anderson (2000), and Coxhead and Jayasuriya (2004).
[3] Technical progress in irrigated rice has in the past been a major source of relative agricultural commodity price changes. These in turn have undoubtedly influenced upland farmers' resource allocation decisions, as discussed above and in Coxhead and Jayasuriya (1994).
[4] APEX shares many features with the well-known ORANI model of the Australian economy (Dixon *et al.*, 1982), but these features have been amended to fit Philippine realities. Input–output data in APEX are drawn from the Philippine Social Accounting Matrix.
[5] More details of the model structure and data can be found in Clarete and Warr (1996) and other papers on the APEX web site, http://rspas.anu.edu.au/economics/povrc.html. Additional description and some illustrative experiments may be found in Warr and Coxhead (1993).
[6] In the decade 1975–1984, the NFA was the second-largest government corporation in terms of these transfers, after the Fertilizer and Pesticide Authority (Intal and Power, 1990: Table 2.2). In 1986–90 it ranked first among subsidy recipients, receiving 2.6 billion pesos (US$100 million), or 24% of all government subsidy expenditures during the period (Philippine Department of Finance, reported in World Bank, 1992:116).

[7] In spite of this limited success, the nominal protection coefficient of maize in the Philippines rose substantially over two decades into the 1990s (Intal and Power, 1990).

[8] Trade volumes can only be fixed by setting some other variable endogenous. We have added a 'shifter' to the import demand equations which, when endogenous, permits us to fix imports without setting endogenous other trade-related variables such as import tariffs.

[9] In the APEX database this sector accounts for about 75% of grain purchases from agriculture.

[10] 'Fixed' means a price remains constant relative to the exchange rate, the *numeraire* price in the model.

[11] One reason is that in order to finance its cereals' purchases and still maintain budget balance, the government must increase revenues, which it does in the model by raising the rate of the lump-sum tax on households. Another reason is that the price supports increase the relative profitability of maize, a crop already supported by trade barriers, and reduce profitability of less protected crops, such as exportables. The price supports thus increase deadweight losses to the economy as productivity of maize production increases.

[12] The 1980 Census of Agriculture reported about 7% of arable land as in fallow.

[13] The Magat and Pantabangan are two of the three major rivers rising in the Cordillera Mountains of Northern Luzon. Cruz *et al.* (1988) provide kilogram-per-hectare estimates for nitrogen loss and urea equivalents; phosphorus loss and solophos equivalents; and potassium loss and potash equivalents. Few differences in soil nutrient content were found across soil types. Corresponding replacement costs for these nutrients were 11.09 pesos, 1.98 pesos and 2.39 pesos per kg, respectively. Note that these estimates, too, are conservative, as they account for losses from sheet erosion only, and neglect changes in economic yield associated with changes in soil structure.

[14] Although we have used the best available estimates of erosion rate and nutrient replacement costs in these calculations, there is still considerable uncertainty about the true values of these parameters. We have recalculated the value of changes in land degradation under alternative parameter values, varying annual erosion losses by 25% and nutrient replacement costs by 50%. The results of this sensitivity analysis are not been reported here due to lack of space, but are reported in Coxhead and Shively (1995). The value of reduced land degradation losses under three alternative closures and nine combinations of parameter values (27 estimates in all) ranges from US$3.9 to 32 million.

References

Balisacan, A. (1992) Parameter estimates of consumer demand systems in the Philippines. APEX Working Paper. Accessed at http://rspas.anu.edu.au/economics/apex/

Clarete, R.L. and Warr, P.G. (1996) The theoretical structure of the APEX model of the Philippine economy. APEX Working Paper. Accessed at http://rspas.anu.edu.au/economics/apex/

Coxhead, I. (1997) Induced innovation and land degradation in developing country agriculture. *Australian Journal of Agricultural and Resource Economics* 41, 305–332.

Coxhead, I. and Jayasuriya, S. (1994) Technical change in agriculture and the rate of land degradation in developing countries: a general equilibrium analysis. *Land Economics* 70, 20–37.

Coxhead, I. and Jayasuriya, S. (1995) Trade and tax policies and the environment in developing countries. *American Journal of Agricultural Economics* 77, 631–644.

Coxhead, I. and Jayasuriya, S. (2004) Development strategy and trade liberalization: implications for poverty and environment in the Philippines. *Environment and Development Economics* 9, 613–644.

Coxhead, I. and Shively, G.E. (1995) Measuring the environmental impacts of economic change: the case of land degradation in Philippine agriculture. University of Wisconsin-Madison, Department of Agricultural Economics: Staff Paper Series No. 384, Madison, Wisconsin.

Coxhead, I.A. and Shively, G.E. (1998) Some economic and environmental implications of technical progress in Philippine corn agriculture: an economy-wide perspective. *Journal of Agricultural Economics and Development* 26, 60–90.

Coxhead, I. and Warr, P.G. (1995) Does technical progress in agriculture alleviate poverty? A Philippine case study. *Australian Journal of Agricultural Economics* 39, 25–54.

Cruz, W., Francisco, H. and Tapawan-Conway, Z. (1988) *The On-site and Downstream Costs of Soil Erosion*. Philippine Institute for Development Studies, PIDS Working Paper No. 88–11, Manila, Philippines.

David, C.C. (2003) Agriculture. In: Balisacan, A.M., and Hill H.(eds) *The Philippine Economy: Development, Policies, and Challenges*, Oxford University Press New York, and Ateneo de Manila University Press, Quezon City, Philippines, pp. 175–218.

David, C.C., Intal, P. and Ponce, E.R. (1992) Organizing for Results: The Philippine Agricultural Sector. Working Paper No. 1992–08. Philippine Institute for Development Studies, Manila, Philippines.

David, W.P. (1988) Soil and water conservation planning: policy issues and recommendations. *Journal of Philippine Development* 15, 47–84.

Dixon, P., Parmenter, B.R., Sutton, J. and Vincent, D.P. (1982) *ORANI: A Multi-sectoral Model of the Australian Economy*. North-Holland, Amsterdam.

Intal, P.S. and Power, J.H. (1990) Trade, Exchange Rate, and Agricultural Pricing Policies: the Philippines. The World Bank, Washington, DC.

Lantican, F.A. and Unnevehr, L.J. (1987) Rice pricing and marketing policy. In: UPLB Agricultural Policy Working Group, *Policy Issues in the Philippine Rice Economy and Agricultural Trade*. University of the Philippines at Los Baños, Center for Policy and Development Studies, College, Laguna, Philippines, pp. 31–72.

National Statistics Office [of the Philippines] (1981) *Census of Agriculture 1980*. NSO, Manila.

Philippine Department of Agriculture (1990) *Philippine Agricultural Development Plan 1991–1995*. Philippine Department of Agriculture, Manila.

Rola, A.C. and Coxhead I. (2003) Does non-farm job growth encourage or retard soil conservation in Philippine uplands *Philippine Journal of Development* 29, 55–84.

Shively, G.E. (1997) Impact of contour hedgerows on upland maize yields in the Philippines. *Agroforestry Systems* 39, 59–71.

Shively, G.E. (1998) Economic policies and the environment: the case of tree planting on low-income farms in the Philippines. *Environment and Development Economics* 3, 83–104.

Shively, G.E., Martinez, E. and Masters, W.A. (2002) Testing the link between public intervention and food price variability: evidence from rice markets in the Philippines. *Pacific Economic Review* 7, 545–554.

Strutt, A. and Anderson, K. (2000) Will trade liberalization harm the environment? The case of Indonesia to 2020. *Environment and Resource Economics* 17(3), 203–232.

Warr, P.G. and Coxhead, I. (1993) The distributional impact of technical change in Philippine agriculture: a general equilibrium analysis. *Food Research Institute Studies* 22, 253–274.

World Bank (1989) *Philippines Natural Resource Management Study*. The World Bank, Washington, DC.

World Bank (1992) *The Philippines: Public Sector Resource Mobilization and Expenditure Management.* World Bank Country Economic Report No. 10056-PH, The World Bank, Washington, DC.

Appendix

Our modelling goal for this chapter is to link changes in important economic parameters to changes in rates of land degradation in critical watersheds. To begin, let T^* be the effective (i.e. quality-adjusted) amount of upland land available for agricultural production. We define this as the product of T, the physical endowment of upland land, and a function of land quality shifters $A(R, M)$, where $R = T_e/T$ is the ratio of land used in an erosive sector to all agricultural land[1] and M is a vector of variables representing land productivity enhancing measures, including technical progress and soil conservation.[2] Given these definitions we can express the demand for effective land by:

$$T^* = T \cdot A(R, M) \tag{6.1}$$

Using subscripts to denote the derivatives of A with respect to R and M, we assume $A_R < 0$; $A_M > 0$; $A_{RR} < 0$ and $A_{MM} < 0$.

We seek an expression for the welfare (or real income) effect of a change in the effective land endowment. Consider a small, open economy in which aggregate expenditures are denoted by $E(P,U) = \min\{P \cdot C \mid U\}$, and aggregate income from the production of goods and services is given by $G(P,V) = \max\{P \cdot Y \mid V\}$, where P, C and Y denote price, consumption and production respectively of a vector of goods, and U is aggregate utility. V is the vector of the economy's effective (i.e. quality-adjusted) factor endowments, of which T is one element. To simplify this heuristic exposition, suppose there are no taxes or subsidies. In equilibrium, aggregate expenditures are equal to aggregate revenues:

$$E(P, U) = G(P, V) \tag{6.2}$$

By totally differentiating (6.2) and collecting terms in prices we obtain an expression for the change in aggregate real income, dY:

$$dY = -H_p dP + G_v dV \tag{6.3}$$

where subscripts on G and E denote partial derivatives of the revenue and expenditure functions with respect to subscripted variables, and dY, the change in real aggregate income, is defined as $dY = E_u dU$, where E_u is the inverse of the marginal utility of income. The term $H_p = (E_p \ G_p)$ is the vector of excess demands for goods; it is positive for net imports and negative for net exports. The derivative $G_v(P,V)$ is the shadow price of the v'th factor, equal to its market price in a competitive economy with no externalities.

Note that from (6.1), a change in effective land area is equal to the weighted sum of changes in the physical land endowment, land use in the erosive activity and management practices:

$$dT^* = AdT + TA_R dR + TA_M dM \qquad (6.4)$$

Substituting (6.4) into (6.3) and for simplicity holding other factor endowments constant yields an expression for the real income impact of price, endowment and factor productivity changes:

$$dY = {\sim}H_p d\boldsymbol{P} + G(AdT + TA_R dR + TA_M dM) \qquad (6.5)$$

Equation (6.5) shows that in the present context four types of change contribute to observed aggregate real income growth: (i) price changes, including the effects of exogenous terms of trade or price policy shocks; (ii) changes in the value of the land endowment due to area growth; (iii) erosion-related changes in agricultural land productivity; and (iv) exogenous increases in land productivity due to technical progress and improved management practices.

Equation (6.5) captures in the simplest way the distinction between models that ignore resource stock depletion and those that account for it. R is a function of relative commodity price changes, which induce shifts in land allocation between more and less erosive crops. If we ignore the costs of resource depletion, $A_R = 0$ by construction. In this case, a shift to more erosive land uses is interpreted as unambiguously welfare-increasing as long as it increases conventional measures of national income. In contrast, with $A_M < 0$, reallocations of land alter average soil fertility, and thus influence real income.

We adapt the APEX model described above to account for changes in soil fertility associated with changes in upland land use and technology. We do this by inferring land degradation costs (rather than attempting to model physical shifts in production functions, which would require rather more data than are presently available). By linking erosion changes to estimated costs of soil nutrient replacement, we construct an estimate of on-site erosion costs associated with changes in (or expansion of) upland activities.

We begin by defining total upland erosion for each of three regions (Luzon, Visayas, Mindanao) as the sum of erosion losses generated in each of twelve agricultural sectors[3]:

$$E_r = \Sigma_i E_{ir}, \qquad i = 1, 2, ..., 12; \quad r = 1, 2, 3 \qquad (6.6)$$

where i is an index of crops and r is an index of regions. Converting (6.6) to proportional changes of variables, we obtain an expression for the total regional change in upland erosion as the share-weighted sum of changes in each sector, the weights being each sector's contribution to total regional erosion in the reference year:

$$e_r = \Sigma_i e_{ir} S_{(i)r}, \qquad i = 1, 2, ..., 12 \; ; r = 1, 2, 3 \qquad (6.7)$$

where $S_{(i)r} = E_{ir}/E_r$, $\Sigma_i\, S_{(i)r} = 1$, and lower-case letters indicate proportional changes of variables, i.e. $e = dE/E$. Each crop grown in each region is associated with a specific rate of soil degradation. When this rate remains constant (in terms of equation (6.4), when $dM = 0$), the proportional change in the amount of erosion produced in sector i is linearly related to the proportional change in land use $(t_{ir}^{(1)})$ in that sector:[4]

$$e_{ir} = t_{ir}^{(1)} \quad i = 1, 2, \ldots, 12; \quad r = 1, 2, 3 \tag{6.8}$$

In the APEX database land and other factors used in agriculture are non-allocable between sectors. Therefore, we cannot read sectoral land use changes directly from the model results. Instead, we calculate them by imputing a land demand function equivalent to the demand functions for allocable inputs in non-agricultural industries. In this expression, at constant relative factor prices the change in land demand is linearly related to changes in sectoral output (written $x_{ir}^{(0)}$). Factor price changes (written $w_{vr}^{(1)}$ for factor v in region r) alter the quantity demanded to an extent governed by the degree of each factor's substitutability with land. Output-augmenting technical progress ($z_{ir}^{(0)}$ for sector i in region r) reduces land demand, other things equal. The imputed land demand for each crop in each region can thus be written in proportional change form as:

$$t_{ir}^{(1)} = x_{ir}^{(0)} + \sum_{v \in A} d_{vir} w(1)vr - z_{ir}^{(0)} \quad i = 1, 2, \ldots, 12; \quad r = 1, 2, 3, \quad v = 1, \ldots, 4 \tag{6.9}$$

where $\sigma_{vir} = \sigma_{ir}^{(1)}\theta_{vir}$ for each non-land agricultural factor v; $\delta_{vtr} = -\sigma_{tr}^{(1)}(1 - q_{vtr})$ for land; $\sigma_{ir}^{(1)}$ is the CES elasticity of factor substitution in sector i and region r, and θ_{vir} is the distributive share of factor v used in agricultural sector i and region r.[5] Land use changes may take place either through changes in derived demand (via changes in sectoral output), or through changes in factor prices (and hence substitution between land and other inputs used in the sector), or via technical progress that reduces land use for a given level of output. If land use in a land-degrading sector does not change then that sector's contribution to degradation remains at its base level.[6]

The national-level change in erosion (e) is computed as the weighted sum of regional erosion changes, where the weights are regional shares of erosion in the reference year:[7]

$$Te = \sum_p U_r e_r \tag{6.10}$$

where $U_r = E_r/E$, and E is the base value of the national aggregate erosion rate, in t per year.

Notes

[1] For simplicity we assume that only one upland crop is erosive; a more general case relates the rate of degradation to the erosivity and area planted of each crop.

[2] Shively (1997) reports productivity impacts of soil conservation on upland maize in the Philippines.

[3] Our analysis focuses on agricultural land degradation, and therefore takes no account of land degradation in mining, forestry or other non-agricultural industries.

[4] For consistency with APEX notation we use superscripts to indicate categories of good. Superscript (0) indicates the supply of a domestically produced good, and (1) the quantity or price of an input to production. Thus $t_{ir}^{(1)}$ denotes the (land) input to production of crop i in region r.

[5] While the structure of the relationships in (6.9) allows for greater flexibility, in practice we use a CES specification as a default, since empirical information about sector-specific factor substitution relationships in agriculture at the required scale of aggregation are lacking.

[6] Total land area in each region is determined elsewhere in the APEX model, so of the 12 land demand relations in each region, only 11 are independent. Land demand in the remaining crop is determined by the condition that the share-weighted sum of land demands be equal to aggregate supply in each region:

$$\sum_i I_{(i)r}^{(1)} t_{ir}^{(1)} = 0 \ \forall r$$

where $I_{(i)r}^{(1)}$ is the initial share of crop i in total land use in region r and

$$\sum_i I_{(i)r}^{(1)} = 1$$

[7] In the absence of better data, our aggregation of erosion losses assumes that a unit of soil lost has the same value in each region.

7 Simulating Soil Erosion and Sediment Yield in Small Upland Watersheds Using the WEPP Model

V. B. ELLA

Land and Water Resources Division, Institute of Agricultural Engineering, College of Engineering and Agro-Industrial Technology, University of the Philippines at Los Baños, College, Laguna 4031, Philippines, e-mail: vbella@up.edu.ph

Introduction

Soil erosion remains one of the most serious environmental problems in developing areas of the tropics. Worldwide, excessive erosion in watersheds causes sedimentation of reservoirs and clogging of irrigation distribution systems, power generation facilities and hydraulic structures resulting in substantial reductions in system and water distribution capacity. In the Philippines, recent estimates show that as much as 95% of the country's 4.7 million ha of watershed area are affected by erosion and that 16 million cubic metres of soil are lost annually (NWRB, 2004). This contributes to heavy sedimentation of the reservoirs of the national irrigation systems in the country, which in turn negatively affects the efficiency of irrigation and hydropower generation systems. Erosion in the Manupali River watershed is typical of that experienced elsewhere.

Apart from reservoir sedimentation and clogging of hydraulic structures, soil erosion also results in soil nutrient depletion or soil fertility reduction with the continuous detachment and transport of nutrient-rich particles from the topsoil. Moreover, soil erosion degrades the quality of downstream surface water resources. Apart from increased turbidity of surface water resources, the eroded sediments may also adsorb and transport agricultural contaminants such as pesticides, phosphates and heavy metals. For watersheds whose run-off waters are consequently discharged into lakes and other aquatic ecosystems, these transported sediments with adsorbed toxic chemicals may pose a serious threat to aquatic life downstream.

Developing procedures for estimating soil loss and sediment yield is an important initial step for gauging the sources, magnitude and importance of soil erosion problems in low-income watersheds, so as to better target policy interventions. Soil loss and sediment yield estimates can serve as a basis for developing soil conservation measures, reservoir sedimentation studies and reservoir design. These estimates can also play an important educational role, both with local communities (see Deutsch and Orprecio, Chapter 3, this volume) and policymakers (see Rola and Coxhead, Chapter 2, this volume).

Soil erosion may be estimated and predicted either through direct measurements or via mathematical and statistical techniques. Both strategies have been employed in the Philippines (e.g. David, 1988; Presbitero et al., 1995; Agua, 1997; Poudel et al., 1999; Cero, 2001; Deutsch et al., 2001; Midmore et al., 2001; Paningbatan, 2001), but to date no study in the Philippines has confronted the problem of estimating soil erosion using advanced computer simulation methods. In fact, with the exception of Paningbatan (2001; see Chapter 8, this volume), who employs a GIS-assisted erosion model, all previous work on erosion estimation in the Philippines has essentially made use of the empirical Universal Soil Loss Equation (USLE), now considered to be largely obsolete.

In recent years the model based on the Water Erosion Prediction Project (WEPP) (Flanagan and Nearing, 1995) has represented a significant advance in methods for combining basic data on the characteristics of a watershed with scientific understanding of soil erosion processes to predict soil erosion.[1] The WEPP model incorporates recent advances in hydrologic, climatologic, soil science, hydraulic and plant research. Unlike the USLE, or even the revised or modified USLE, the WEPP model is capable of simulating or predicting soil loss and sediment yield at a watershed scale. The model has been validated using numerous data sets and is now considered the state-of-the-art tool for predicting soil erosion at a catchment scale. Although the WEPP model was developed for use in the USA, the ultimate goal of its developers was more general applicability (Flanagan and Nearing, 1995), and it has proved to be applicable worldwide. In Asia, it has been applied in China (Wu et al., 2002) but the study reported here is the first application in the Philippines of which the author is aware.

The goal of this chapter is to provide some evidence regarding soil erosion in the Manupali River watershed. This is done by using the WEPP model to estimate and predict soil erosion and sediment yield in subwatersheds of the main Manupali River watershed system. Specifically, this chapter outlines the development of hillslope and watershed-scale WEPP erosion simulation models for selected small upland subwatersheds of the Manupali River. Soil erosion and sediment yield predictions are presented for selected subwatersheds. In addition, the WEPP model is used to study changes in erosion and sediment yield that have arisen due to changes in land cover and land management practices in the selected subwatersheds. This chapter provides an overall assessment of erosion risk in the Manupali River watershed and demonstrates the strong linkage between land cover – as determined by smallholders operating in the watershed – and environmental outcomes. It provides an underpinning for the simulation work undertaken by Shively and Zelek (Chapter 10, this volume) and provides a perspective on erosion outcomes derived from a formal model as a complement to the farmer-collected data on water quality presented by Deutsch and Orprecio (Chapter 3, this volume) and the simulation results presented by Paningbatan (Chapter 8, this volume).

Overview of the WEPP Model

The WEPP model is a process-based, continuous computer simulation model for predicting water-induced soil erosion either on a hillslope or at the scale of a watershed. The model is based on the fundamental principles of stochastic weather generation, infiltration theory, hydrology, soil physics, plant science, hydraulics and mechanics of soil erosion and sediment transport. The model consists of nine conceptual components corresponding to climate generation, hydrology, soils, irrigation, plant growth, residue decomposition, hydraulics of overland flow, erosion and (where applicable) winter processes. Figure 7.1 provides a simplified representation of the model. An appendix to this chapter presents additional detail on the model as relevant to the intended application.

Description of the Test Watersheds

The watersheds under examination are tributaries of the Manupali River basin (see Plate 4). As described in Chapter 1, the Manupali River is a major irrigation water source for the Manupali River Irrigation System (MANRIS) with a

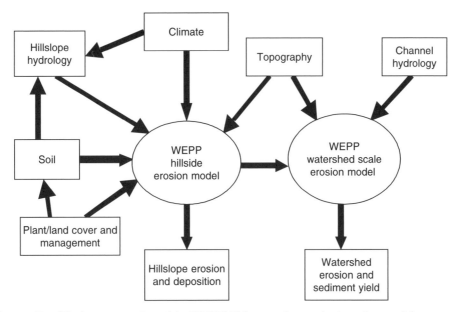

Fig. 7.1. Simplified representation of the WEPP hillslope and watershed erosion model.

service area of 4000 ha. It also drains into the Pulangi River, a major waterway in Mindanao, and feeds a series of major hydroelectric power generation plants located on that river.

This study uses data from four tributaries of the Manupali River: the Tugasan, Maagnao, Alanib and Kulasihan rivers.[2] These smaller rivers drain into the Manupali River from its northern side (see Plate 4). Their topography is rolling to hilly, and ranges in elevation from about 400 m above sea level at the outlet of the Kulasihan River to about 2938 m at the upstream peak shared by the Maagnao and Alanib watersheds. Soils in these subwatersheds are predominantly clayey due to the extent of fine-grained volcanic rocks, various sedimentary derivatives and pyroclastics (BSWM, 1985). Annual rainfall over the watersheds averages 2470 mm. Rainfall peaks in June to October and declines in November to February with March and April being the driest months. Mean monthly temperature ranges from 30°C to 34°C. Relative humidity averages 80% for elevations over 500 m and 60–75% for elevations less than 500 m. The existing land cover is comprised of about 29% dense forest, 35% agricultural crops (predominantly maize and vegetables) and 17% shrubs and small trees (Bin, 1994). Physiographic details of these subwatersheds are provided in Table 7.1.

Methods

Application of the WEPP model is based on a large body of relevant physical data collected previously in the selected watersheds. These data include climatologic, hydrologic, soils, topographic, land use, vegetation and cover management practices. Data and information were gathered from projects undertaken in the SANREM study as well as from agencies of the Philippine government (especially the Bureau of Soils and Water Management, (BSWM), and the Philippine Meteorological Agency, PAGASA). The collected secondary data were compiled and checked for adequacy and reliability. Field visits were conducted to perform some ground checking of the selected watersheds as an aid in validating the secondary data and information gathered. The coordinates of the rainfall and streamflow gauging stations were verified using a global positioning system (GPS) unit.

Table 7.1. Physiographic details of the test watersheds.

Name of watershed	Watershed outlet		Range of elevation (m)	Area (ha)
	Latitude	Longitude		
Tugasan	8°02′N	124°52′E	1134 to 2742	5024
Maagnao	8°03′N	124°55′E	1200 to 2938	2397
Alanib	8°01′N	124°59′E	750 to 2938	7513
Kulasihan	8°01′N	125°05′E	400 to 2278	10,042

Climatological data were pre-processed prior to disaggregating them into a format acceptable to the WEPP model. Original spreadsheet programmes were developed for climatological data processing. Processed data were then used as input to the Breakpoint Climate Data Generator (BPCDG) (Zeleke *et al.*, 1999) to form part of the input files for the WEPP model. Separate hill-slope models were developed for various portions of the subwatersheds based on WEPP using average soil, topographic and land cover conditions. Watershed-scale models were then developed, using the hillslope models and other watershed characteristics. Both hillslope and watershed models were then used for simulating surface run-off, soil erosion and sediment yield on a continuous (i.e. instantaneous) mode.

One drawback of the current version of the WEPP model is that it can be applied only to relatively small watersheds. The current version of the model restricts the number of channels per watershed to fewer than 30. The larger watersheds (Kulasihan and Alanib) were not modelled because initial attempts showed that the number of channels to be interconnected exceeded this limit. As a result, efforts were focused on applying the model to the Maagnao and Tugasan watersheds, the smallest of the four subwatersheds originally considered for the study.

For each of these subwatersheds, both hillslope and watershed models were developed using the input data prepared in appropriate format. Topographic maps were used to properly define the channel configuration and the slope. In the absence of complete land cover data, assumptions were made on the most likely crop cover conditions based on data presented by Bin (as reported in Coxhead and Buenavista, 2001) and from data obtained from BSWM (1985). Soils data and slope data were both incorporated in the hill-slopes and channel models.

Watershed-scale WEPP erosion models for the Maagnao and Tugasan sub-watersheds were developed using the hillslope models and relevant watershed characteristics. The WEPP watershed model for the Maagnao River watershed consisted of 10 channels and 22 hillslopes. The WEPP watershed model for the Tugasan River watershed consisted of 15 channels and 35 hillslopes. The WEPP watershed models developed for the Maagnao and Tugasan watersheds are capable of generating streamflow and sediment discharge at the outlet of the watershed on a monthly and yearly basis. The WEPP models are also capable of generating a sediment delivery ratio and soil loss figure for each hillslope on each side of the channel, as well as other relevant parameters.

In the absence of a complete and accurate set of data on sediment discharge and short-increment streamflow, the WEPP watershed models were further refined by repeated trial simulations until realistic values of sediment load were obtained. This route was chosen instead of the usual model calibration and validation procedure. The latter was infeasible given the data limitations.

In view of data limitations, hillslope models were developed for the upstream, midstream and downstream zones of each subwatershed based on average soil, topographic and crop cover conditions. To establish a land cover baseline, the upstream and midstream hillslope models were assumed to be covered with small trees, brushes and grasses. The downstream hillslopes were assumed to be covered by agricultural crops and grasses. These baseline cropping patterns were derived from BSWM data for Bukidnon province. Simulations then focused on examining the impact of varying land cover in each zone, in line with the historical, policy-induced land cover changes, such as those highlighted by Coxhead and Shively (Chapter 1, this volume).

Results and Discussion

The WEPP hillslope models were executed on a continuous mode using break-point climate data obtained at the Alanib weather station for the calendar year 2000. The results of hillslope model simulation are shown in Table 7.2. Simulation results for the downstream portions of the watershed are consistent with the measured data in previous studies at the site (e.g. Midmore et al., 2001). WEPP model simulations indicate that the soil loss and sediment yield in the upstream and midstream hillslopes proved to be much lower than those in the downstream hillslopes. This pattern reflects the impact of greater soil disturbances in the downstream hillslopes due to more intensive agricultural practices in the downstream portions of the watershed. In the absence of dis-aggregated spatial data on observed soil losses, however, it is difficult to assess the overall accuracy of the simulated values.

Note that in interpreting the results in Table 7.2 one might conclude that a parameter with a unit of t/ha is ten times a parameter with a unit of kg/m^2 (since 1 tonne is 1000 kg and 1 ha is 10,000 m^2), but it should be stressed that sediment yield (t/ha) will not always be the same as soil loss (kg/m^2). For hillslope level simulations, sediment yield is the amount of soil loss per unit area that reaches the most downstream portion of the hillslope. Soil loss, on the other hand, is the total amount of soil lost per unit area over the entire hillslope. With soil deposition, the sediment yield will be lower than the soil loss. This is the reason why the sediment yield for the downstream hillslope (approximately 6.1–11.1 kg/ha) is lower than the soil loss (approximately 18.1–38.6 kg/ha). Conversely, the sediment yield is not 181–386 t/ha but only 60.7–110.9 t/ha because of soil deposition. This is plausible since, while a large amount of soil can be lost (detached from the soil) in cultivated areas in the downstream hillslope class, soil deposition may also take place in the cultivated portion for a number of reasons. For one, soil disturbances may indi-

Table 7.2. WEPP hillslope model simulation results.

Hillslope	Soil loss (kg/m^2)	Sediment yield (t/ha)
Upstream (brush/grass/trees)	0.125 to 0.127	1.25 to 1.27
Midstream (brush/grass)	0.125 to 0.134	1.25 to 1.34
Downstream (cabbage/maize/potato/ sugarcane/grass)	18.1 to 38.6	60.7 to 110.9

rectly pave the way for the creation of some soil sinks or trapping mechanism for soils being transported over the surface. This is particularly true if tillage takes place along contours. Moreover, soil disturbances may also lead to increased surface roughness and hence greater soil surface resistance to soil transport thereby increasing deposition. However, this is not to say that soil disturbances are a good thing, since despite the deposition that takes place, the magnitude of soil particles reaching the extreme end of the hillslope per unit area (i.e. the sediment yield) under cultivated or disturbed conditions is still much higher than in the case of no cultivation because of greater sources for sediment yield (i.e. greater soil loss in the former than in the latter). This further explains why the sediment yield for the downstream hillslopes (60.7–110.9 t/ha) is much higher than those for the upstream and midstream hillslopes (1.25–1.34 t/ha), despite deposition.

Sample watershed-scale model simulation results for the Maagnao River subwatershed are shown in Figs 7.2 and 7.3 for streamflow and sediment discharge, respectively, using breakpoint climate data obtained from the Alanib weather station for the year 2000. While instantaneous comparisons between simulated and observed values of both surface run-off and sediment yield proved to be infeasible due to the absence of observed instantaneous hydrograph and sediment discharge data, model simulation yielded reasonable values. For instance, while the WEPP watershed model appears to over-predict the streamflow in Fig. 7.2, it should be emphasized that the observed values were obtained from one-time measurement each month of the year, and therefore did not fully account for 'unobserved' high run-off events. The same is true for the sediment discharge data. As a result, simulated values appear to be higher than the observed values because the former more completely account for all run-off events. Nevertheless, simulation results indicate that the

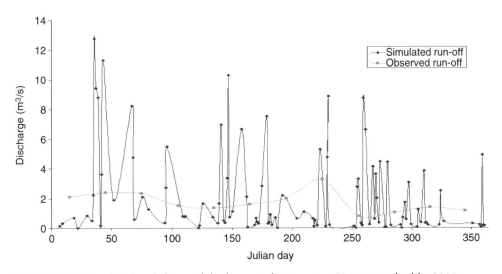

Fig. 7.2. WEPP-simulated and observed discharge at the Maagnao River watershed for 2000.

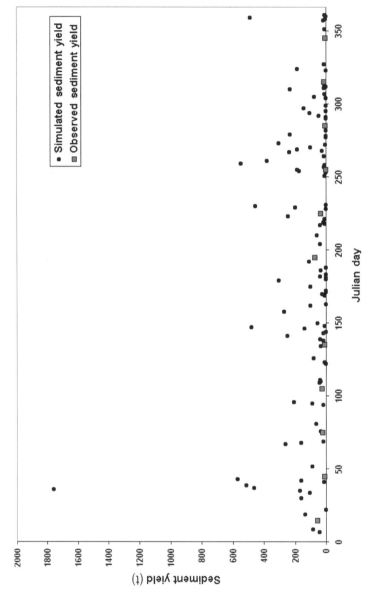

Fig. 7.3. WEPP-simulated and observed sediment yield in the Maagnao River watershed for 2000.

watershed-scale WEPP model is capable of generating realistic instantaneous hydrograph and sediment yield values.

Using the WEPP model for the Maagnao River watershed, further analysis focused on examining the possible effects of changes in land cover conditions on soil erosion and sediment yield. This part of the modelling elucidates the potential for using the adapted WEPP model to study land use management (or mismanagement). The model is capable of providing quantitative predictions of the soil erosion consequences of any human-induced activity at the site for the given soil, slope and climatic conditions. For this study climatological breakpoint data for year 2000 (observed at the Alanib weather station) were used to simulate soil erosion and sediment yields under several hypothetical land use scenarios. These scenarios corresponded to varying the cultivated area of the watershed at discrete points between 0% and 100%. Table 7.3 shows the results of the model simulations in terms of sediment yield. Simulation results indicate that, in general, increasing the percentage of cultivated or cropped area in the watershed increases sediment yield. For instance, when the entire watershed is not cultivated and instead is covered by small trees and grasses, the sediment yield is 1.9 t/ha/year. If 20% of the watershed area is cultivated and planted to maize, sugarcane and vegetables (with grasses in between), the sediment yield according to model simulation is increased to 11.1 t/ha/year. On the other hand, if half of the entire watershed is cultivated and planted to the same set of crops, the sediment yield would increase substantially to 19.9 t/ha/year. Further increases in cultivated area to 75% and 100% of watershed area result in sediment yields of 24.9 and 48.5 t/ha/year, respectively. Plots of these results, as shown in Figs 7.4 and 7.5, further indicate that increasing the extent of soil disturbances by agricultural cultivation generates a non-linear increase in soil loss and sediment yield that is more pronounced as watershed integrity becomes more greatly compromised. For example, as cropped area doubles from 10% of watershed area to 20%, sediment yield increases by less than a factor of two; but as cropped area doubles from 50% of watershed area to 100%, sediment yield more than doubles. This suggests that outcomes become more sensitive to changes in land cover as human disturbance of watershed cover increases.

Table 7.3. Watershed erosion prediction under varying land cover conditions.

Watershed area cultivated and cropped (%)	Sediment yield (t/ha/yr)
0	1.9
10	8.8
20	11.1
50	19.9
75	24.9
100	48.5

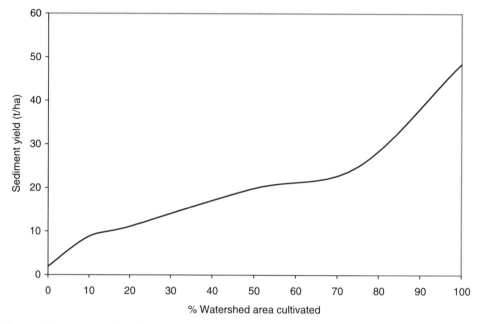

Fig. 7.4. Effect of agricultural area expansion on sediment yield in the Maagnao River watershed.

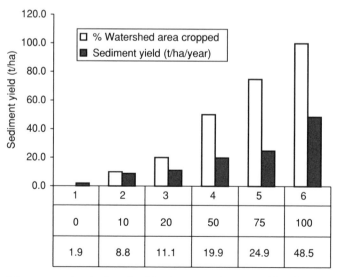

Fig. 7.5. Effect of the extent of agricultural cultivation on sediment yield in the Maagnao River watershed.

Summary and Conclusions

This chapter reported results from computer-based soil erosion models, developed using WEPP. The models were used to simulate soil erosion and sediment yield in selected small upland watersheds in the Manupali River basin in the Philippines. Hillslope models were developed based on WEPP using average soil, topographic and crop cover conditions at various portions of the selected test watersheds of the river basin. WEPP watershed-scale models were also developed for the small subwatersheds using hillslopes models and relevant watershed characteristics. Model simulations under continuous mode using breakpoint climatological data provided predictions of erosion and sediment yield within the typical range of values for croplands and rangelands. Simulated sediment yield generated by upstream (brush/grass/trees), midstream (brush/grass) and downstream (cabbage/maize/potato/sugarcane/grass) hillslope WEPP models ranged from 1.25 to 1.27, 1.25 to 1.34 and 60.7 to 110.9 t/ha/year, respectively. Results for the downstream hillslopes proved to be consistent with measured data from previous studies at the site. Simulated surface run-off and sediment yield generated on a continuous basis by the WEPP watershed model also generally agreed reasonably with observed data.

The watershed-scale simulations demonstrated the capability of the WEPP model for predicting the relative effects of varying land cover conditions on sediment yield. While the applicability of the WEPP model at the chosen site in particular and in the Philippines in general is constrained to a large extent by the lack of necessary input data (including continuous rainfall observations, accurate and updated land cover data, information on agricultural management practices, spatial variability of soil properties, and channel hydraulic characteristics) the results are nevertheless useful in providing information regarding physical processes in the watershed. Although model validation has been constrained by the availability of accurately measured continuous hydrograph and sediment discharge data, this study has nevertheless generated customized mathematical modelling tools based on WEPP for soil erosion prediction on a hillslope or watershed scale. Over time, and with additional data, these modelling approaches may be further improved. In the mean time, these results fill a gap in knowledge regarding the magnitude of erosion losses and sediment yields in the Manupali River watershed. As such, they provide a useful starting point for economy–environment modelling, such as that undertaken by Shively and Zelek (Chapter 10, this volume). The results also can serve to foster local and national policy discussions leading to the formulation of policies concerning land cover management in upland watersheds in the Philippines.

Acknowledgements

The author wishes to thank NSERL, USDA-ARS for the WEPP model, Dr William Deutsch and Heifer International Philippines for providing much of the needed data, and Dr Vel Suminguit and the SANREM field staff for additional data. This study was conducted as part of the USAID-funded project under the auspices of the SANREM-CRSP Environmental Research Grant coordinated by DOST-PCARRD.

Appendix

This appendix presents the essential features of the WEPP model, as relevant to the application of the model to the Philippine watersheds discussed in the text. For additional information interested readers should consult Flanagan and Nearing (1995), who provide a detailed description of the WEPP model.

Climate component

The climate component of WEPP (Nicks et al., 1995) makes use of either measured or synthetic climatic data. The climate generator component determines the number and distribution of precipitation events using a two-state Markov chain model. A rainfall disaggregation model using a double exponential function is also incorporated in the climate component to generate time–rainfall breakpoint data from daily values. All equations used in the climate component of the WEPP model are reported by Nicks et al. (1995). The climatological parameters including disaggregated rainfall are used in estimating peak rate, duration and total amount of run-off.

Hydrology component

WEPP's hydrology component (Gilley and Weltz, 1995; Savabi and Williams, 1995; Savabi et al. 1995; Stone et al., 1995) includes simulation of infiltration, surface run-off, soil evaporation, plant transpiration, percolation and subsurface flow. Infiltration calculation is based on the Green-Ampt-Mein-Larson equation (Mein and Larson, 1973) and a ponding time calculation for unsteady rainfall (Chu, 1978) based on cumulative rainfall, rainfall intensity, saturated hydraulic conductivity, matric potential and soil moisture deficit. Equations for the simulation of infiltration, ponding time and excess rainfall can be found in Stone et al. (1995). The effect of spatial and temporal variability of soil water intake is simulated by adjusting the effective hydraulic conductivity of the soil due to human disturbances such as tillage and natural phenomena such as soil surface sealing and vegetal roots. Equations that account for these variabilities can be found in Alberts et al. (1995).

Excess rainfall is consequently routed downslope by the WEPP model to estimate the overland flow hydrograph using a kinematic wave approach. The effect of surface depressional storage on surface run-off is simulated in the WEPP model. Peak run-off is calculated using either a semi-analytical solution of the kinematic wave equation in the case of single storm event simulation or an approximation of the kinematic wave model based on soil, rainfall, slope and friction factors in the case of continuous simulation (Stone *et al.*, 1995). The kinematic wave model is of the form:

$$\frac{\partial h}{\partial t} + \frac{\partial q}{\partial x} = v \qquad (7.1)$$

where h represents flow depth, q represents unit discharge, x is distance and t is time.

Solution methodologies for estimating peak flows using equation (7.1) and equations for run-off duration are presented in Stone *et al.* (1995). Peak run-off and run-off duration are consequently used by the WEPP model in calculating shear stress, sediment transport capacity and rill erosion.

The water balance and percolation component of the WEPP model is based on a water balance model developed by Williams and Nicks (1985). Water redistribution in the soil profile is done by WEPP based on simulations of evapotranspiration and soil moisture routing. Details of the water balance procedures and equations can be found in Savabi and Williams (1995).

Plant growth component

The plant growth component of the WEPP model (Arnold *et al.*, 1995) simulates the plant and residue status above and below the soil surface. The crop growth model is essentially based on the EPIC model (Williams *et al.*, 1984). The model assumes phenological crop development based on daily accumulated heat units and a harvest index for partitioning grain yield. Both cropland and rangeland plant growth can be dealt with by this component. Canopy cover and height, biomass above and below ground surface, leaf area index, root growth and basal area are estimated on a daily basis. Mathematical details of the plant growth component can be found in Arnold *et al.* (1995).

Soil component

The soil component of WEPP (Alberts *et al.*, 1995) simulates the temporal variability of the various soil properties that influence the soil erosion process. It maintains a daily accounting of the status of the soil and surface variables which include random roughness, ridge height, bulk density, saturated hydraulic conductivity, rill and inter-rill erodibility parameters and critical hydraulic shear stress. The soil component also accounts for the effect of tillage, consolidation and rainfall on soil and surface variables. Tillage effects are expressed in terms of bulk density, roughness, ridge height and residue cover in the soil component. Baseline rill soil erodibility and critical hydraulic shear values for a freshly tilled condition are adjusted to other conditions by consolidation because of wetting and drying. The mathematical details of the soil component can be found in Alberts *et al.* (1995).

Management component

The effect of the various land management practices on hydrology and erosion for a given site is simulated by the WEPP model through its management component. This component determines the changes in soil physical properties, surface roughness and cover conditions due to such practices as tillage, crop harvest, grazing and other management activities.

Erosion component

The WEPP model is fundamentally based on the concept that soil erosion is a process of detachment and transport. Both hillslope erosion and channel erosion are simulated by the WEPP model. The WEPP hillslope erosion model (Nearing *et al.*, 1989; Foster *et al.*, 1995) is based on a steady-state sediment continuity equation of the form:

$$\frac{\partial G}{\partial x} = D_f + D_i \tag{7.2}$$

where x is distance downslope (in metres), G is the sediment load (kg/s/m), D_f is the rill erosion rate (kg/s/m^2) and D_i represents the inter-rill erosion rate (kg/s/m^2).

Rill detachment is calculated for the case when the hydraulic shear stress exceeds the critical shear stress of the soil and when sediment load is less than the sediment transport capacity as:

$$D_f = D_c\left(1 - \frac{G}{T_c}\right) \tag{7.3}$$

where D_c is the detachment capacity by rill flow and T_c is the sediment transport capacity in the rill. When the hydraulic shear stress exceeds the critical shear, the detachment capacity is calculated as:

$$D_c = K_r\left(\tau_f - \tau_c\right) \tag{7.4}$$

where K_r is the rill erodibility parameter, τ_f represents the flow shear stress acting on the soil and τ_c is the critical shear stress of the soil.

Net deposition is calculated for the case when the sediment load is greater than the sediment transport capacity as:

$$D_f = \frac{\beta V_f}{q}(T_c - G) \tag{7.5}$$

where V_f is the effective fall velocity for the sediment, q is the unit discharge and β is a raindrop-induced turbulence coefficient.

Inter-rill erosion, on the other hand, is treated as independent of distance x. Inter-rill sediment delivery is calculated as a function of the square of the rainfall intensity, inter-rill erodibility and factors representing the ground and canopy cover effects.

The flow shear stress is calculated as a function of the average slope gradient and the Darcy-Weisbach friction factor of the rill. The sediment transport capacity at the end of the slope is calculated based on a modified Yalin sediment transport equation (Yalin, 1963). Other mathematical details of the hillslope erosion component of the WEPP model can be found in Foster *et al.* (1995).

The WEPP's channel hydrology and erosion model (Ascough *et al.*, 1995, 1997) is also included for watershed scale simulations. Sediment load in the channel is calculated as a function of the incoming upstream load from hillslopes, channels and impoundments; the incoming lateral load from adjacent hillslopes and impoundments; and detachment of bed materials. Like the hillslope erosion model, the sediment load is based on a quasi-steady state sediment continuity equation similar in form to equation (7.2) with the right hand side replaced by lateral sediment inflow and detachment or deposition by flow, respectively. Hydraulic shear stress is a function of friction slope, calculated based on spatially varied flow assumption. Sediment transport capacity is calculated based on the Yalin sediment transport equation. The mathematical details of the channel hydrology and erosion model can be found in Ascough *et al.* (1995, 1997).

Notes

[1] The WEPP model was developed at the National Soil Erosion Research Laboratory (NSERL) of the Agricultural Research Service, United States Department of Agriculture in collaboration with the Natural Resource Conservation Service (NRCS), Forest Service, Bureau of Land Management, and in cooperation with more than ten universities and numerous international scientists (Laflen *et al.*, 1997).

[2] Editors' note: these are also the tributaries studied by Deutsch and Orprecio (Chapter 3, this volume).

References

Agua, M.M. (1997) Predicting peak rates of run-off and soil loss from a watershed. PhD dissertation, University of the Philippines at Los Baños, College, Laguna, Philippines.

Alberts, E.E., Nearing, M.A., Weltz, M.A., Risse, L.M., Pierson, F.B., Zhang, X.C., Laflen J.M. and Simanton J.R. (1995) Soil component. In: Flanagan, D.C. and Nearing, M.A. (eds) *USDA-Water Erosion Prediction Project: Hillslope Profile and Watershed Model Documentation*. NSERL Report No. 10. USDA-ARS National Soil Erosion Research Laboratory, West Lafayette, Indiana, pp. 7.1–7.47.

Arnold, J.G., Weltz, M.A., Alberts E.E. and Flanagan, D.C. (1995) Plant growth component. In: Flanagan, D.C. and Nearing, M.A. (eds) *USDA-Water Erosion Prediction Project: Hillslope Profile and Watershed Model Documentation*. NSERL Report No. 10. USDA-ARS National Soil Erosion Research Laboratory, West Lafayette, Indiana, pp. 8.1–8.41.

Ascough, J.C., Baffaut, C.B., Nearing, M.A. and Flanagan, D.C. (1995) Watershed model channel hydrology and erosion processes. In: Flanagan, D.C. and Nearing, M.A. (eds) *USDA-Water Erosion Prediction Project: Hillslope Profile and Watershed Model Documentation.* NSERL Report No. 10. USDA-ARS National Soil Erosion Research Laboratory, West Lafayette, Indiana, pp. 13.1–13.20.

Ascough, J.C., Baffaut, C.B. Nearing, M.A. and Liu, B.Y. (1997) The WEPP watershed model: I Hydrology and erosion. *Transactions of the ASAE* 40, 921–933.

Bin, L. (1994) The impact assessment of land use change in the watershed area using remote sensing and GIS: a case study of Manupali watershed, the Philippines. Unpublished Master's thesis, Asian Institute of Technology, Bangkok, Thailand.

Bureau of Soils and Water Management, Department of Agriculture (BSWM) (1985) Land resources evaluation report for Bukidnon province: the physical land resources volume 1. BSWM-DA, Manila, Philippines.

Cero, L.L. (2001) Erosion, sediment yield and water quality assessment in a subwatershed of Taal Lake, Philippines. PhD Dissertation, University of the Philippines at Los Baños, College, Laguna, Philippines.

Chu, S.T. (1978) Infiltration during unsteady rain. *Water Resources Research* 14, 461–466.

Coxhead, I. and Buenavista, G. (eds) (2001) *Seeking Sustainability: Challenges for Agricultural Development and Environmental Management in a Philippine Watershed.* Philippine Council for Agriculture, Forestry and Natural Resources Research and Development, Department of Science and Technology, Los Baños, Laguna, Philippines.

David, W.P. (1988) Soil and water conservation planning: Policy issues and recommendations. *Philippine Journal of Development Studies* XV, 47–84.

Deutsch, W.G., Busby, A.L., Orprecio, J.L., Labis, J.B. and Cequiña, E.Y. (2001) Community-based water quality monitoring: from data collection to sustainable management of water resources. In: Coxhead, I. and Buenavista, G. (eds) *Seeking Sustainability: Challenges for Agricultural Development and Environmental Management in a Philippine Watershed.* Philippine Council for Agriculture, Forestry and Natural Resources Research and Development, Department of Science and Technology, Los Baños, Laguna, Philippines, pp. 138–160.

Flanagan, D.C. and Nearing, M.A. (eds) (1995) USDA water erosion prediction project (WEPP). NSERL Report No.10. USDA-ARS National Soil Erosion Research Laboratory, West Lafayette, Indiana.

Foster, G.R., Flanagan, D.C., Nearing, M.A., Lane, L.J., Risse, L.M. and Finkner, S.C. (1995) Hillslope erosion component. In: Flanagan, D.C. and Nearing, M.A. (eds) *USDA Water Erosion Prediction Project: Hillslope Profile and Watershed Model Documentation.* NSERL Report No. 10. USDA-ARS National Soil Erosion Research Laboratory, West Lafayette, Indiana, pp. 11.1–11.12.

Gilley, J.E. and Weltz, M.A. (1995) Hydraulics of overland flow. In: Flanagan, D.C. and Nearing, M.A. (eds) *USDA-Water Erosion Prediction Project: Hillslope Profile and Watershed Model Documentation.* NSERL Report No. 10. USDA-ARS National Soil Erosion Research Laboratory, West Lafayette, Indiana, pp. 10.1–10.7

Laflen, J.M., Elliot, W.J., Flanagan, D.C., Meyer, C.R. and Nearing, M.A. (1997) WEPP-predicting water erosion using a process-based model. *Journal of Soil and Water Conservation* 52, 96–102.

Mein, R.G. and Larson, C.L. (1973) Modeling infiltration during a steady rain. *Water Resources Research* 9, 384–394.

Midmore, D.J., Nissen, T.M. and Poudel, D.D. (2001) Making a living out of agriculture: some reflections on vegetable production systems in the Manupali watershed. In: Coxhead, I. and Buenavista, G. (eds)

Seeking Sustainability: Challenges for Agricultural Development and Environmental Management in a Philippine Watershed. Philippine Council for Agriculture, Forestry and Natural Resources Research and Development, Department of Science and Technology, Los Baños, Laguna, Philippines, pp. 94–111.

National Water Resources Board (NWRB) (2004) Water for food: Aiming for self-sufficiency and rural development. Available at: http://www.nwrb.gov.ph Accessed June 10, 2004.

Nearing, M.A., Foster, G.R., Lane, L.J. and Finkner, S.C. (1989) A process-based soil erosion model for USDA-water erosion prediction project technology. *Transactions of the ASAE* 32, 1587–1593.

Nicks, A.D., Lane, L.J. and Gander, G.A. (1995) Weather generator. In: Flanagan, D.C. and Nearing, M.A. (eds) *USDA-Water Erosion Prediction Project: Hillslope Profile and Watershed Model Documentation.* NSERL Report No. 10. USDA-ARS National Soil Erosion Research Laboratory, West Lafayette, Indiana, pp. 2.1–2.22.

Paningbatan, E.P. (2001) Geographic information system-assisted dynamic modeling of soil erosion and hydrologic processes at a watershed scale. *Philippine Agricultural Scientist* 84, 388–393.

Poudel, D.D., Midmore, D.J. and West, L.T. (1999) Erosion and productivity of vegetable systems on sloping volcanic ash-derived Philippine soils. *Soil Science Society of America Journal* 63, 1366–1376.

Presbitero, A.L., Escalante, M.C., Rose, C.W., Coughlan, K.J. and Ciesiolka, C.A. (1995) Erodibility evaluation and the effect of land management practices on soil erosion from steep slopes in Leyte, the Philippines. *Soil Technology* 8, 205–213.

Savabi, M.R. and Williams, J.R. (1995) Water balance and percolation. In: Flanagan, D.C. and Nearing, M.A. (eds) *USDA-*

Water Erosion Prediction Project: Hillslope Profile and Watershed Model Documentation. NSERL Report No. 10. USDA-ARS National Soil Erosion Research Laboratory. West Lafayette, Indiana, pp. 5.1–5.14.

Savabi, M.R., Skaggs, R.W. and Onstad, C.A. (1995) Subsurface hydrology. In: Flanagan, D.C. and Nearing, M.A. (eds) *USDA-Water Erosion Prediction Project: Hillslope Profile and Watershed Model Documentation.* NSERL Report No. 10. USDA-ARS National Soil Erosion Research Laboratory, West Lafayette, Indiana, pp. 6.1–6.14.

Stone, J.J., Lane, L.J., Shirley, E.D. and Hernandez, M. (1995) Hillslope surface hydrology. In: Flanagan, D.C. and Nearing, M.A. (eds) *USDA-Water Erosion Prediction Project: Hillslope Profile and Watershed Model Documentation.* NSERL Report No. 10. USDA-ARS National Soil Erosion Research Laboratory, West Lafayette, Indiana, pp. 4.1–4.20.

Williams, J.R. and Nicks, A.D. (1985) SWRRB: A simulator for water resources in rural basins: An overview. In: D.G. DeCoursey (ed.) *Proceedings of Natural Resources Modeling Symposium,* Pingree Park, Colorado, October 16–21, 1983, pp. 17–24. USDA-ARS. ARS-30.

Williams, J.R., Renard, K.G. and Dyke, P.T. (1984) EPIC: a new model for assessing erosion's effect on soil productivity. *Journal of Soil and Water Conservation* 38, 381–383.

Wu, J.Q., Xu, A.C. and Elliot, W.J. (2002) Adapting WEPP for forest watershed erosion modeling. Paper presented at 12th ISCO Conference, Beijing.

Yalin, M.S. (1963) An expression for bed-load transportation. *Journal of Hydraulic Division, American Society of Civil Engineers* 98, 221–250.

Zeleke, G., Winter, T. and Flanagan, D. (1999) BPCDG: Breakpoint Climate Data Generator for WEPP using observed standard weather data sets. USDA-ARS National Soil Erosion Research Laboratory, West Lafayette, Indiana.

8 Identifying Soil Erosion Hotspots in the Manupali River Watershed

E. P. PANINGBATAN JR

Department of Soil Science, University of the Philippines at Los Baños, College, Laguna 4031, The Philippines, e-mail: dss@laguna.net

Introduction

Soil erosion is a major threat to poorly managed watersheds. This is particularly evident in sloping uplands where residents utilize a significant part of a watershed for crop cultivation and where other land uses such as roads, housing and recreation facilities also pose an erosion risk. With intense monsoon rainfall, the disturbance of erodible soils on steep slopes leads to soil erosion. Unless conservation-oriented land management practices are employed, patterns of land use typically found in watersheds such as the Manupali River watershed will generate substantial soil erosion (e.g. Deutsch and Orprecio, Chapter 3; Ella, Chapter 7, this volume). Erosion, in turn, will continue to degrade the upland environment, and in the longer run may worsen the poverty of upland farmers as well as generate downstream costs.

Historically, the management of watershed resources has not been a goal in many watersheds, including the Manupali River watershed (see Rola and Coxhead, Chapter 2, this volume). But increasingly, as policy attention turns to protection and management of watershed resources, a need has emerged to better understand the impact of land use on hydrology and soil erosion at a watershed scale. With appropriate methodology the on-site effect of soil erosion and run-off on soil productivity and the off-site effects on water quality and quantity can be better understood and evaluated. There is also a need for decision support tools capable of locating erosion-prone areas within a large watershed so that interventions or mitigating measures can be properly formulated and targeted. These tools can aid cost-effective erosion minimization in areas where mitigation costs are typically measured in terms of reductions in the incomes of extremely poor households. In the context of the Manupali River watershed, in particular, where financial and technical resources are scarce, the reality of economic and biophysical conditions dictate that efforts to alter current land uses must be targeted where they will be most cost effective.

©CAB International 2005. *Land Use Change in Tropical Watersheds: Evidence, Causes and Remedies* (eds I. Coxhead and G. Shively)

This chapter discusses the use of a geographic information systems (GIS) assisted soil erosion model to identify and map erosion locations within the Manupali River watershed that are particularly susceptible to erosion risk due to physical characteristics and current uses. The chapter may be viewed as complementary to Chapter 7. The goal here is to go beyond the prediction of overall soil losses associated with various land uses and to identify erosion hotspots, that is, locations where either changes in land use or more stringent soil conservation efforts should be undertaken in order to protect the watershed. Such findings may serve to help target efforts directed at influencing land uses or cropping practices, such as those identified by Midmore *et al.* in Chapter 9.

Methods

The analysis of this chapter relies on predictions from the Predicting Catchment Run-off and Soil Erosion for Sustainability (PCARES) soil erosion simulation model. PCARES was developed and validated using data from the Mapawa and Alanib catchments of the Manupali River watershed. PCARES is a GIS-assisted physical model that simulates run-off and soil erosion of a catchment area during a rainfall event. The model predicts the spatial and temporal distribution of soil erosion rates, and can be used as a decision support tool to identify erosion hotspots as influenced by soil, slope and land use/cover of the watershed. It can also predict sediment and water discharge rates at the outlet of a catchment area, thereby quantifying the off-site effects of watershed soil erosion.

The model uses PCRaster, a GIS software package capable of cartographic and dynamic modelling. The method allows simulation of the hydrologic and sediment transport processes occurring on a three-dimensional landscape. The concept, structure and script can be understood by a researcher with basic expertise in programming. Paningbatan (2001) describes the theoretical basis, computer programming and validation of the model. The basic input parameters to run the model include digitized maps of elevation, land use, soil, rain station and a time series rainfall amount. The hydrology component of the model involves the calculation of water flow velocity using the widely applied Manning's equation which allows for calculation of discharge rates at shorter time intervals than do other methods. The model incorporates a system of calculating sediment transport as described by Rose and Freebairn (1985).

A wide range of data is required to apply the model. In the case of the Manupali River watershed, data sets were developed from GIS, remote sensing (RS) and global positioning systems (GPS), and these were used to construct a series of geo-referenced thematic maps. GIS-supported maps of land use, elevation and soil are key ingredients in the list of essential parameters required to generate a spatial distribution of erosion in a watershed. Georeferenced data on biophysical resources in the watershed used to construct the model included vegetation types, road locations, river and stream locations, and land attributes such as elevation, slope and soil type. Key components are described below.

Digital elevation model (DEM) map

A DEM map was prepared by digitizing a 1:50,000 scale topographic map with contour intervals of 20 m. This was converted into a raster map called the DEM map (see Plate 6) using the GIS software packages SURFER and PCRaster. The DEM map was then used to delineate the sub-catchment areas of the watershed, as presented in Plate 4. The sub-catchment areas of the major tributaries of the Manupali River are listed in Table 8.1. For purposes of calibrating the PCARES model, the total watershed area was assumed to be 38,369 ha.

The DEM map was also used to generate a slope map, shown in Plate 7, which was used to derive the total areas under each slope category that are presented in Table 8.2. Approximately 57% of the total area of Manupali River watershed has a slope greater than 18%. These slopes, if cultivated with no provision for effective soil conservation measures, are sites of intense soil erosion.

Soil types

A soil map was prepared by digitizing soil types as identified in an existing reconnaissance soil survey report from the province of Bukidnon. Six soil

Table 8.1. Major sub-catchments in the Manupali River watershed.

Sub-catchments	Area (ha)	% of total
Alanib	7,530	20
Maagnao	2,412	6
Panabo-Tugasan	5,209	14
Timago	3,718	10
Others	19,500	51
Total	38,369	100

Table 8.2. Area under various slope categories in Manupali River watershed.

Per cent slope	Area (ha)	% of total
0–3	798	2
3–8	4,348	11
8–18	11,599	30
18–25	7,690	20
> 25	13,934	37
Total	38,369	100

types had been identified in the watershed, with an additional category of undifferentiated forest soil. A map of soil types (*see* Plate 8) closely aligns with a map of elevation. Overall areas of each soil type are presented in Table 8.3. The texture of the soil is clay, except for a very small area of silt loam (less than 1% of the watershed) that belongs to the San Manuel soil series. The clay soils that belong to soil order *Ultisol* have relatively low soil erodibility if in their natural, highly aggregated state due to a very high cohesive strength and very high infiltration rates. However, once the soil aggregates are disturbed and compacted during cultivation, these soils become highly erodible.

Land uses

A map of major land uses in the watershed is presented in Plate 9. The land use map resulted from image processing and classification based on the RS image of LandSat 7 taken in 2002. For this study the image was classified into forest, cultivated upland, cultivated lowland, roads and rivers. Those areas classified as forest and cultivated uplands occupied 52% and 41%, respectively, as shown in Table 8.4. The cultivated uplands are potential sites of intense soil erosion if not protected with sufficient amount of plant cover. Practically no accelerated soil erosion occurs under good forest vegetation. Roads and footpaths are also sites with very high soil erosion potential (Paningbatan, 2001).

Table 8.3. Major soil types in the Manupali River watershed.

Soil type	Area (ha)	% of total
Adtuyon clay	15,380	40
Kidapawan clay	10,750	28
Forest soil	11,420	30
Maapag clay	377	1
Alimodian clay	330	<1
San Manuel silt loam	340	<1
Total	38,369	100

Table 8.4. Major land uses affecting soil erosion in the Manupali River watershed.

Land use	Area (ha)	% of total
Forest	20,056	52
Cultivated lowland	356	<1
Cultivated upland	15,730	41
Road	369	<1
Rivers, streams	1,858	5
Total	38,369	100

Simulating Soil Erosion Hotspots

The PCARES model, constructed and calibrated as described above, was used to simulate a rainfall event and thereby identify a series of simulated soil erosion outcomes in the Manupali River watershed. In addition to GIS-assisted thematic maps just discussed, a rainfall hydrograph (Fig. 8.1) was constructed based on rainfall data gathered using an automatic rain gauge installed inside the watershed. An observed rainfall event was linked to measurements of major stream flow and soil erosion in Mapawa creek, a sub-catchment of the Manupali River watershed located in barangay Songko (see Plate 2). A series of parametric erosion coefficients associated with the types of land use presented in Table 8.5 completes the data input requirements of PCARES.

Plate 10 shows the geo-referenced map of predicted soil erosion in the watershed, an output of the simulated rainfall event. The soil erosion hotspots

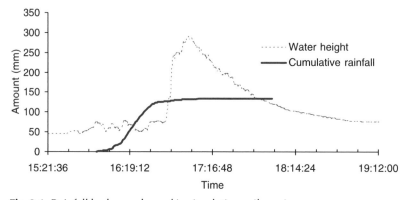

Fig. 8.1. Rainfall hydrograph used in simulating soil erosion.

Table 8.5. Input values for soil surface cover, Mannings cofficient, saturated soil infiltration rate of the different land use used to run the model PCARES.

Land use	Surface cover (%)	Manning's *n* (m³/s)	Saturated infiltration (mm/h)
Fallow	95	0.12	280
Road	05	0.025	2
Footpath	05	0.03	3
Stream	05	0.05	15
Tributary	10	0.04	15
Riparian buffer	85	0.1	100
Cultivated (maize, vegetables)	20	0.06	17
Cultivated (with conservation)	48	0.1	60
Tree plantation	85	0.1	100

Source: Paningbatan (2001).

are indicated in red. These are distributed mainly in the cultivated uplands with steep slopes. In total, the simulation identifies erosion hotspots covering approximately 5269 ha (Table 8.6), or roughly 14% of the watershed area. These sites are those that should be targeted for stringent soil conservation efforts or restrictions on highly erosive practices. In contrast, sites with very low soil erosion rates are in the forestlands, covering 53% or 20,210 ha of the watershed. In these areas efforts should focus on maintaining the current state of vegetative cover in order to protect the watershed.

Summary and Conclusions

This chapter has summarized the use of a GIS-assisted model to simulate soil erosion at a watershed scale during a run-off-producing rainfall event. The simulation model was used to locate erosion hotspots in the Manupali River watershed. The basic inputs to the model include raster maps of the elevation, slope, soil and land use of the catchment area, and a time series of rainfall events.

The simulation model was parameterized to represent the Manupali River watershed in the Philippine province of Bukidnon. Land use maps showing forests, cultivated upland areas, cultivated lowland areas, roads and rivers were prepared using a geo-referenced LandSat 7 image taken in 2002. Digital elevation and slope maps were generated from existing topographic maps with a scale of 1:50,000 and a 20-m contour interval. A soil map was digitized from existing reconnaissance soil maps of the study area.

The model identified the location of soil erosion hotspots. These are areas of high erosion potential, distributed, not surprisingly, mainly in the cultivated uplands with steep slopes. As defined here, erosion hotspots constitute around 5269 ha or 14% of the watershed area. Stringent soil conservation efforts should be undertaken in these erosion hotspots while the current state of vegetative cover in the forestlands should be maintained in order to protect the watershed. Moreover, attempts to alter cultivated practices, along the lines suggested by experimental results of Midmore *et al.* (Chapter 9, this volume) should focus on these areas. Another use for these data would be to help inform local and provincial officers charged with

Table 8.6. Land area with various soil erosion intensities in the Manupali River watershed.

Soil erosion	Area (ha)	% of total
Very low	20,210	53
Low	10,310	27
Hotspots	5,269	14
River	2,570	7
Total	38,359	100

Source: Simulation model results.

responsibility for protecting and rehabilitating key areas of the watershed. Such efforts might involve restrictions on certain types of crops or cropping practices (such as those examined by Shively and Zelek, Chapter 10, this volume) or might include efforts to modify land use through initiatives linked to providing payments for environmental services (as described by Pagiola *et al.* Chapter 11, this volume). Regardless of the approach, the main message that emerges from this analysis is that, in terms of reducing physical rates of erosion, returns to efforts will vary considerably across the landscape and will be greatest in areas identified here as hot spots. The next obvious step would be to assess the overall economic costs and potential benefits of targeting efforts in such areas.

References

Paningbatan, E.P. (2001) GIS-assisted dynamic modeling of soil erosion and hydrologic processes at watershed scale. *The Philippine Agricultural Scientist* 84, 388–393.

Rose, C.W. and Freebairn, D.M. (1985) A new mathematical model of soil erosion and deposition processes with application to field data. In: El Swaify, S.A., Moldenhauer, W.C. and Lo, A. (eds) *Soil Erosion and Conservation*. Soil Conservation Society of America, Ankeny, Iowa.

Plate 1.

Plate 2.

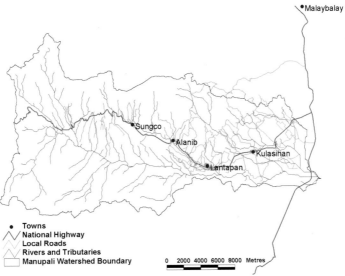

Plate 1. Map of the Philippines showing the province of Bukidnon (see Chapter 1).
Source: http://encyclopedia.thefreedictionary.com/

Plate 2. Map of the Manupali River watershed showing major towns, roads and tributaries (see Chapters 1 and 8). Source: Ian Coxhead, University of Wisconsin-Madison.

Plate 3.

Plate 4.

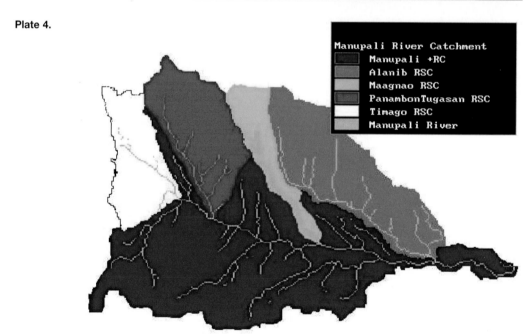

Manupali River Catchment
Manupali +RC
Alanib RSC
Maagnao RSC
PanambonTugasan RSC
Timago RSC
Manupali River

Plate 3. Aerial photograph of the Manupali River watershed (see Chapter 1).
Source: SANREM project archives, University of Georgia.

Plate 4. Map of subwatersheds and catchments (see Chapters 1, 3, 7 and 8).
Source: Eduardo Paningbatan, University of the Philippines Los Baños.

Plate 5.

Plate 6.

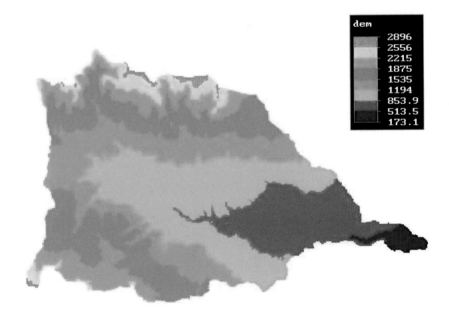

Plate 5. Photograph of *Tigbantay Wahig* (water watch) team members at work (see Chapter 3).
Source: Bill Deutsch, Auburn University.

Plate 6. Elevation map of the Manupali River watershed (see Chapter 8).
Source: Eduardo Paningbatan, University of the Philippines Los Baños.

Plate 7.

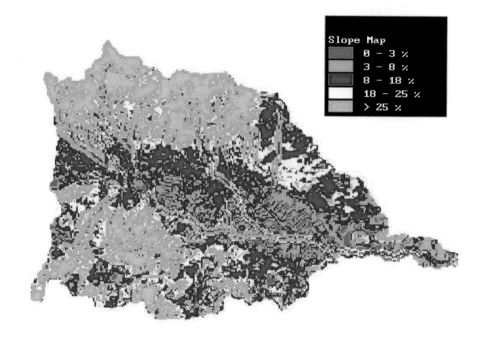

Slope Map
- 0 – 3 %
- 3 – 8 %
- 8 – 18 %
- 18 – 25 %
- > 25 %

Plate 8.

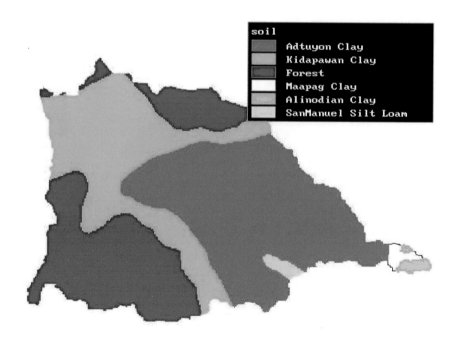

soil
- Adtuyon Clay
- Kidapawan Clay
- Forest
- Maapag Clay
- Alinodian Clay
- SanManuel Silt Loam

Plate 7. Slope map of the Manupali River watershed (see Chapter 8).
Source: Eduardo Paningbatan, University of the Philippines Los Baños.

Plate 8. Soils map of the Manupali River watershed (see Chapter 8).
Source: Eduardo Paningbatan, University of the Philippines Los Baños.

Plate 9.

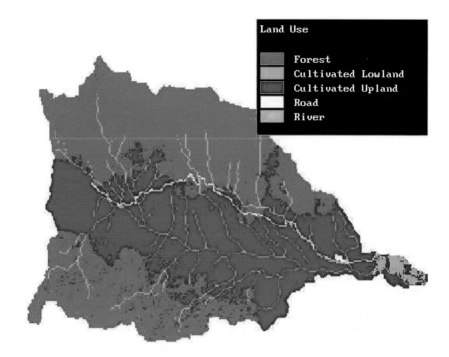

Plate 10.

Plate 9. Land use map of the Manupali River watershed (see Chapter 8).
Source: Eduardo Paningbatan, University of the Philippines Los Baños.

Plate 10. Map of the Manupali River watershed showing erosion hotspots (see Chapter 8).
Source: Eduardo Paningbatan, University of the Philippines Los Baños.

Plate 11.

Plate 12.

Plate 11. Photograph showing typical agricultural fields in the Manupali River watershed (see Chapter 9).
Source: SANREM project archives, University of Georgia.

Plate 12. Photograph showing typical vegetable plots in the Manupali River watershed (see Chapter 9).
Source: SANREM project archives, University of Georgia.

9

Alternatives to Traditional Annual Crop Agriculture in the Uplands

Biophysical Evidence from the Manupali River Watershed

D. J. MIDMORE,[1] D. D. POUDEL,[2] T. M. NISSEN,[3*] A. DAÑO[4] AND G. ZHU[5]

[1]School of Biological and Environmental Sciences, Central Queensland University, Rockhampton, QLD 4702, Australia, e-mail: d.midmore@cqu.edu.au; [2]Department of Renewable Resources, University of Louisiana, Lafayette, LA 70504, USA; e-mail: ddpoudel@louisiana.edu; [3]Bureau of Economic and Business Affairs, US Department of State, Washington, DC 20520, USA; e-mail: nissentm@state.gov,e-mail:erdb@laguna.net; [4]Department of Environment and Natural Resources, Los Baños, The Philippines, e-mail: erdb@laguna.net; [5]School of Biological and Environmental Sciences, Central Queensland University, Rockhampton, QLD 4702, Australia, e-mail: g.zhu@cqu.edu.au

Introduction

Maize and commercial vegetable production are important features of the agricultural landscape in the Manupali River watershed.[1] These crops, as currently produced, are associated with a number of environmental concerns in the watershed, including high rates of soil erosion (see Chapters 7 and 8, this volume) and – in the case of vegetables – high rates of pesticide use. Soil erosion and associated declines in land productivity and farm income in steep-land crop production systems have been shown to be major threats to agricultural sustainability in the Philippines (Presbitero et al., 1995; Poudel et al., 1999, 2000) and elsewhere (Alegre and Rao, 1996; Gachene et al., 1997; Roose and Ndayizigiye, 1997; Sen et al., 1997). In addition to the on-site effects arising from soil erosion, a number of off-site effects have received attention as detailed elsewhere in this book. The sedimentation of dams, reservoirs, irrigation canals, and degradation of the quality of coastal habitats and tourist

*The views expressed are the author's own and not necessarily those of the US Department of State or the US Government.

locations reflect some of the downstream negative impacts of soil erosion that have major ecological, economic and environmental consequences (Heusch, 1993; Ciesiolka *et al.*, 1995).

A number of field research and computer modelling studies have sought to identify appropriate land management techniques that minimize soil erosion and nutrient losses and improve land productivity of steep-lands (Presbitero *et al.*, 1995; Alegre and Rao, 1996; Gachene *et al.*, 1997; Poudel *et al.*, 1999, 2000). As an example, Poudel *et al.* (1999) studied the effects of contour hedgerows, strip cropping and contour planting on soil erosion on steep-land vegetable systems and reported that contour hedgerows reduce soil erosion by 30% compared to the conventional practice of planting vegetables up-and-down the slope. Other researchers (Tacio, 1993; Comia *et al.*, 1994) have reported even greater effectiveness of contour hedgerows on soil erosion control. Alley cropping (Comia *et al.*, 1994; Paningbatan *et al.*, 1995) and several agroforestry systems (Tacio, 1993; Nautiyal *et al.*, 1998) have also been found to reduce soil erosion and increase agricultural productivity and farm profitability. However, farmers in sloping lands have not often adopted these erosion-control techniques for various reasons including: the capital and labour costs of installing hedgerows; loss of field space occupied by hedgerows (Shively, 1997; Cramb *et al.*, 1999); and the focus on N-fixing trees and shrubs when nitrogen has not been limiting, and on high input systems where additional fertilizer can mask the effects of loss of nutrients. Steep-land vegetable farms operate on small landholdings and often suffer total crop failure due to pest damage or drought (Poudel *et al.*, 1998). Steep-lands have an increased risk of soil erosion, declining soil productivity and catastrophic income loss compared with less steeply sloping lands (El-Swaify, 1997; Shively, 2001). On steep fields, appropriate soil conservation practices, cropping sequences and plant protection measures are necessary to enhance the sustainability of agricultural production systems (Poudel *et al.*, 1998). One of the most widely accepted techniques to improve economic sustainability is a diversified production system in which one crop may partially compensate for the failure of another. Intercropping provides an opportunity for diversification and more efficient use of light, soil nutrients and water (Midmore, 1993).

Trees are often promoted as alternatives to annual crops. Indeed, during the past decade many farmers on Mindanao, including farmers in the Manupali River watershed, have begun to plant timber trees for income generation. Nevertheless, the effectiveness of young trees in controlling erosion in a plantation or intercropping system has not been closely studied. Intercropping full-canopy timber trees and annual crops in a commercial vegetable production system is a relatively new concept, and understanding of the effectiveness of such systems in soil and water conservation and enhancing land productivity is limited. Among the few reported studies to date, Nissen *et al.* (2001) studied tree–vegetable intercropping in the Philippines and reported that intercropping becomes increasingly attractive to farmers as labour becomes scarce relative to land, as the need to minimize cash inputs becomes more important, and as trees increase in value relative to annual crops. The study did not, however, consider the impacts of tree–intercrop systems on soil erosion, nor did it examine the effectiveness (in terms of reduced erosion and enhanced soil productivity) of allowing land to remain fallow between vegetable cropping events. Although fallow rotations are frequently employed by farmers, fallows are rarely managed to encourage the growth of preferred species and most evidence regarding the effectiveness of fallow rotations is anecdotal.

To address these research gaps, the current study was designed to compare annual crop systems with and without trees, and to monitor tillage-induced and rainfall-induced soil erosion from steep-land vegetable systems. In this chapter we report research results based on field experiments designed to evaluate tree-based (*Eucalyptus grandis*) and improved fallow (sunflower, *Tithonia diversifolia*) production systems with annual crops to determine the effectiveness of such systems to minimize soil erosion and enhance land productivity. The trials ran from 1998 to 2003 in the Manupali River watershed. We show that both sunflower and trees alone were effective in minimizing soil erosion. But tree–annual crop intercropping did not minimize tillage-induced and water-related soil loss until the natural vegetation was allowed to grow once cropping was stopped. Tillage-induced erosion was the greatest source of erosion over the study period, with the exception of a few intense rain-storms which resulted in substantial soil losses. After rotation with sunflower, soil pH, N%, C% and Mg^{2+} were all raised, and annual crop yields were greater than on fields with no rotation or liming (a practice that is common on local farms). Yields of annual crops with lime were mostly greater than those without, but, if planted with trees, yields of annual crops were reduced to zero by the fourth cropping season. Our data provide a basis for an understanding of tree species interactions and fallow systems with annual crops, and their effects on soil nutrients. Our results suggest that development of tree–annual crop production and management strategies could be a key component in the design of sustainable steepland annual crop production systems.

Materials and Methods

Data used in this analysis were collected as part of a research study conducted from 1998 to 2003 in Victory, Lantapan, Bukidnon, in the Manupali River watershed (124°47′ to 125°08′ E and 7°57′ to 8°08′ N) at 1210 m above sea level in Mindanao, the Philippines. The soil parent materials were thick deposits of volcanic ejecta either deposited in place or transported from upslope as col-luvial or alluvial materials. In 1995 we established earlier field experiments at the site (Poudel *et al.*, 1999). These experiments had 12 treatments (three crop sequences × four management practices) with 8 m wide and 19 m long erosion–run-off plots laid out in a completely randomized block design with two blocks. The average slope of these erosion–run-off plots was 42%. Each plot was separated by galvanized iron sheets 22.5 cm above and 20 cm below the ground surface, and each plot had a soil collection buffer at the bottom of the plot. In 1998, at the beginning of the current study, a new field experiment was initiated and superimposed. This experiment had five randomly assigned treatments and a varying number of replications (unbalanced design) on the original research plots (Table 9.1). The five experimental treatments consisted of: (i) sunflower rotation for seven seasons, then reverting to cropping; (ii) trees alone; (iii) trees with lime application and annual crops for five seasons; (iv) annual crops and lime; and (v) annual crops without lime. Details of the sequence of annual crops are presented in Table 9.2. Seedlings of approximately 6 months of age were acquired locally.

Data from the field experiment were collected over ten seasons, from May 1998 to December 2003. During and at the end of each cropping season, bordered sections from each plot with annual crops were harvested, and the produce was weighed fresh before marketing. Eroded soils were collected manually from soil collection buffers after every rain event, air-dried and weighed. Run-off was also quantified, through collection in a graded weir system. Collected sediment was also air-dried and weighed. Tillage-induced eroded soil (gravitational soil loss) was collected from soil collection buffers whenever there was a significant amount of soil movement, air-dried and weighed.

Tree measurements were based on diameter at breast height (DBH). The heights of individual trees were measured on a regular basis throughout the experiment. Tree diameter was measured using a measuring tape and height was measured using a 5 m graduated pole until height exceeded that. Stand basal area (SBA) was calculated as:

Table 9.1. Treatments and their description for the soil erosion study (unbalanced completely randomized design, Mindanao, the Philippines).

Treatments	Description
1. Sunflower	Eight plots; sunflower (*Tithonia diversifolia*) established from introduced plant stems to the soil surface; no fertilizer or pesticides applied; stems and leaves removed after season 7 and plots recultivated with tomato, maize, beans or cabbage.
2. Trees only	Six plots; 6-month old *Eucalyptus grandis* seedlings planted 2000 trees/ha in 5 m spaced rows across the slope and 1 m between plants; no fertilization and minimal weeding to 50 cm from the stem during the first season only.
3. Tree/annual crop/lime	Two plots; 6-month old *E. grandis* seedlings planted as in treatment 2; annual crops (tomato, maize, beans or cabbage) planted up-and-down the slope. Standard management practices of fertilizer application and pesticides application as necessary done for tomato, maize and cabbage crops and reported in Poudel *et al.*, (1999); lime applied at the rate of 2 t/ha in May 1998.
4. Annual crop/lime	Six plots; tomato, maize, beans or cabbage planted in strip plots (×2), on the contour (×2), between hedgerows (×1) and up-and-down the slope (×1); lime applied at the rate of 2 t/ha in May 1998. Vegetable crops and maize managed as for treatment 3.
5. Annual crop/no lime	Two plots; tomato, maize, beans and cabbage planted up-and-down the slope and between hedgerows. Standard fertilization and pesticide application procedures followed.

Table 9.2. Comparisons of crop yield (t/ha) with different production systems and lime treatments.

Season	Crop	Rainfall (mm)	No lime	With lime	Trees and lime	Following sunflower
				Treatments		
1	Maize (incl. cobs)	965	4.43[b]	9.07[a]	7.85[a]	–
2	Cabbage	665	7.73[b]	16.97[a]	4.55[b]	–
3	Dry beans	809	0.36[b]	0.76[a]	0.19[c]	–
4	Cabbage	938	0.04[b]	1.83[a]	0	–
5	Dry beans	1091	0	0.34	0	–
6	Maize (grain)	1306	0.14[a]	0.36[a]	–	–
7	Cabbage	529	0	0	–	–
8	Tomato	978	8.00[a]	6.40[b]	–	7.61[a]
9	Beans	n/a	0	1.77[a]	–	4.54[b]
10	Maize	1055	0.77[b]	2.87[a]	–	3.14[a]
11	Cabbage	413	0[a]	4.88[a]	–	7.39[a]
12	Tomato	1155	0	0	–	0

Note: Values within a row with a different letter are significantly different at $P < 0.05$ (season 2 at $p < 0.15$).

$$SBA = (\pi/40000) \times DBH^2 \times density$$

where SBA is the stand basal area in m²/ha, DBH is the arithmetic mean of diameter at breast height of trees in the plot, and density is the number of trees per hectare.

Soil sampling was conducted as follows. Random surface (0–15 cm) composite soil samples were collected (six from each plot) after the seventh season of the trial sequence. Samples were analysed for soil organic matter, total N, available P, Ca, Mg and K. Organic matter was determined by a modified Walkley-Black method (Nelson and Sommers, 1982), while total N was determined by a modified Kjeldahl method (Black, 1965). Available P was determined with Bray-2 extraction (Murphy and Riley, 1962). Exchangeable Ca, Mg and K were extracted with NH_4OAc at pH 7, and the cations in the leachate were measured by atomic absorption spectrophotometry. The infiltration rate was determined using a single ring infiltrometer and potable water on two occasions for cultivated plots.

Statistical Analyses

Our statistical results rely on one-way ANOVA tests to quantify the differences between the treatments for the range of quantified parameters. Changes in infiltration rate and selected soil characteristics were determined by a student's t-test. Unbalanced two-way analysis of variance with interaction was undertaken by using GLM procedure of SAS (SAS, 2000) to test for any seasonal effect on soil erosion. The treatment effects were determined by the Waller-Duncan Multiple Range Test.

Results and Discussion

Soil erosion

Soil detachment and subsequent recovery was recorded in eight out of the ten crop seasons. As Figs 9.1 and 9.2 indicate, treatments had significant effects on total soil loss, rainfall-related loss and tillage-induced soil loss (i.e. the difference between total and rainfall-related soil loss). The seasonal effect was also significant, and largely due to the relationship between in-crop rainfall and soil loss (Fig. 9.3).

It is of particular interest that tillage-induced soil erosion in some of the earlier seasons was almost double that of soil loss attributed to rainfall (see Figs 9.1 and 9.2). Two cropping seasons out of the ten had only tillage-induced soil erosion, while one cropping season had less than 1 t/ha of soil erosion from run-off from all plots. Only in the final cropping season in which erosion was measured was there a relatively small amount of tillage-induced soil erosion compared to soil loss from run-off. The greater amount of soil erosion from

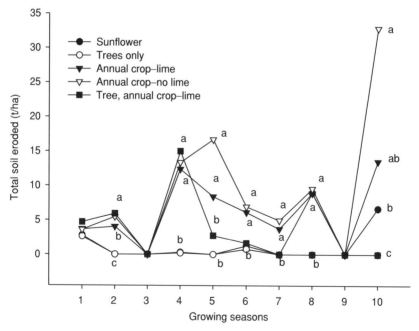

Fig. 9.1. Relationship between total eroded soil in a sequence of cropping including with annual crops, a sunflower rotation and trees with or without annual crops.
Note: Values within a growing season with a different letter are significantly different at $P < 0.05$.

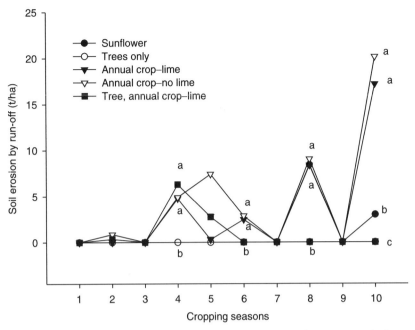

Fig. 9.2. Relationship between eroded soil by run-off in a sequence of cropping including with annual crops, a sunflower rotation and trees with or without annual crops.
Note: Values within a growing season with a different letter are significantly different at $P < 0.05$.

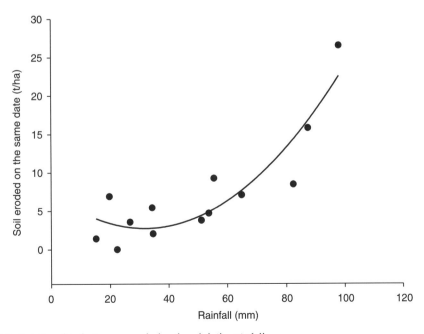

Fig. 9.3. Relationship between eroded soil and daily rainfall.

run-off during the last cropping season was due to the coincidence of land cultivation and strong rain events during that season.

From the beginning of the experiment, both sunflower and trees alone were effective in minimizing soil erosion, whereas erosion was evident in cultivated annual crop plots with or without trees. Plots with sunflower or trees alone developed a weed surface cover (around sunflower plants and between trees) that minimized erosion. Evidently, if cultivation and up-down annual cropping continues beneath trees, the trees have minimal benefit in containing soil erosion. During the fifth season, erosion associated with rainfall was reduced in the tree–annual crop plots compared to annual crops alone (trees had a stand basal area of 2.3 m^2/ha), and in the sixth season onwards it was minimized since no annual crop was planted. Trees without the weed cover and when young were not effective at reducing erosion, when combined with annual crops: even during the fourth season after tree planting there was no difference in eroded soil between them and the limed annual crop control. In the fifth season, tillage erosion was responsible for soil loss in the tree–annual crop plots, but thereafter, when annual crops were not planted and weeds were allowed to grow between trees, erosion was minimized and was equivalent to that from plots with trees alone. The erosion in tree–annual crop plots from the sixth season onwards mimicked those of trees alone: weed growth, predominately the grass *Paspalum conjugatum*, under the tree canopy was effective in controlling erosion. Similar findings showing that surface cover management, even under a tree canopy, is essential for erosion control were reported by Hashim *et al.* (1995) from their soil erosion study in Malaysia.

From the eighth season and subsequently, erosion measurements in continuously cropped plots and those that were returned to cropping after a seven-season rotation with sunflower showed no significant differences in quantity of eroded soil; only in the final season for which erosion was measured was there a difference between the two.

Soils and infiltration

Data on soil physical and chemical parameters were collected immediately after harvest of the sunflower rotational crop. The sunflower rotation significantly raised soil pH, N%, C% and Mg^{2+} by comparison with the continuously cropped plots, and although Ca^{2+} was greater than in other plots, the difference was not significant (Table 9.3). It is possible that the sunflower recycled more deeply available Ca^{2+} and Mg^{2+}. Perennials are known to effect nutrient pumping from reserves deep in the soil (Schroth and Lehmann, 2003), but no data to determine the rooting depth of sunflower were collected in the current experiment.

Rates of infiltration were only recorded on cultivated plots. Infiltration responded positively to the inclusion of sunflower as a fallow crop (Table 9.4). However, after fallow plots were re-cultivated, the improvement was only notable at the top of each plot, for at the lower end of each plot differences were not significant (averaging 100 cm/h in 2001 and 81 cm/h in 2002).

The effects of in-crop practices to contain erosion (e.g. contour planting, hedgerow and strip cropping) were also negligible at down-slope positions (Poudel *et al.*, 1999), as dislodged soil was shifted down the slope and ensured good filtration where it was detained.

Tree growth

Tree growth, quantified as diameter at breast height (DBH), stand basal area and plant height (data not shown for the latter two), showed no response to previous cropping pattern. Nor, for the first eight cropping seasons, were these measures sensitive to whether trees were planted in plots with or without associated annual crops (see data for DBH in Fig. 9.4). This result contrasts with data from earlier trials, in which a marked growth benefit, believed to be due to capture by trees of surplus nutrients supplied to annual crops, was noted for trees planted together with annual crops (Nissen *et al.*, 2001). Several explanations are possible. There may have been a tree-species effect: the relatively shallow-rooting *Parasenthianthes falcataria* responded positively to intercropping,

Table 9.3. Comparison of soil chemical properties (mean ±SE) with different cropping systems (May 2001).

Treatments	pH	N (%)	K (ppm)	C (%)	Ca (Meq/ 100g)	Mg (Meq/ 100g)
Sunflower	5.02[a] (0.07)	0.60[a] (0.02)	207.38 (30.33)	5.32[a] (0.24)	2.34 (0.29)	1.19[a] (0.18)
Annual crop/lime	4.68[b] (0.05)	0.52[b] (0.02)	284.75 (26.71)	4.06[b] (0.22)	2.17 (0.19)	0.55[b] (0.12)
Annual crop/ no lime	4.71[b] (0.11)	0.53[b] (0.00)	309.00 (13.50)	3.57[b] (0.25)	2.05 (0.97)	0.37[b] (0.25)
Trees, annual crop/lime	4.60[b] (0.03)	0.47[b] (0.09)	206.25 (23.25)	4.30[ab] (0.66)	1.43 (0.21)	0.50[b] (0.05)
df	13	12	14	14	14	14
P	0.005	0.041	0.159	0.005	0.508	0.026

Note: Values within a column with a different letter are significantly different at *P* < 0.05.

Table 9.4. Soil infiltration (cm/h) as affected by rotation and lime application in 2001 and 2002.

	May 2001	May 2002
After sunflower	111[a]	67
Annual crop/lime	92[a]	37
Annual crop/no lime	40[b]	33

Note: Values within a row with a different letter are significantly different at *P* < 0.05.

whereas the deeply tap-rooting *Eucalyptus grandis* did not. Additionally, with 4 years of high-input cultivation prior to the start of this experiment, soil in all plots would have been expected to be of reasonable fertility down the profile, or with such low soil pH (see Table 9.3) that the growth of trees was limited with or without the extra fertilizer applied to the trees planted with annual crops. The latter is unlikely since trees with annual crops also received a single lime application and would therefore be expected to have responded favourably early in their growth if soil pH in the plots without annual crops was so low as to constrain growth. Data on soil pH collected 3 years after the application showed no residual effect of the liming (see Table 9.3). The positive response of tree growth during the final 2 years (see Fig. 9.4) may be due to a leaching of Ca^{2+} deeper in the soil profile to a depth where tree roots were concentrated.

Annual crop yields

It was not possible to determine whether there was a trend for annual crop yields to decline over time in all plots (see Table 9.2), for between-season comparisons were confounded by the planting of different species. With the exception of one season, yields of annual crops with lime exceeded those without lime across seasons, but if planted with trees, yields of annual crops were reduced to zero by the fourth cropping season. The yield decline in annual crops with trees relative

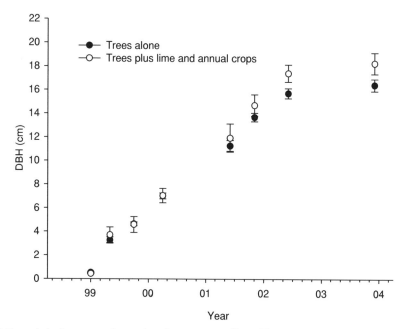

Fig. 9.4. Tree girth diameter at breast height (DBH) as affected by planting alone or with liming and intercropping of annual crops for the first 2 years.

to that of the sole annual crop yields was closely related to the increase in tree stand basal area.[2] Similar relationships have been reported for other tree species at this location (Nissen and Midmore, 2002). The negative effect of *E. grandis* on annual crop yields in this series of experiments was more severe than that of other tree species. For example, the reduction in annual crop yield was $70\%/m^2$ stand basal area, compared with $8-18\%/m^2$ stand basal area for *Paraserianthes falcataria* and *E. deglupta* (Nissen and Midmore, 2002).

Following return of sunflower rotation plots to cropping in the eighth season, yields of beans showed a positive response to the fallow but those of tomato did not. In the ninth season, maize responded favourably to both lime and sunflower rotation. While there was no response of soil pH to liming measured at 0–15 cm depth (see Table 9.3), there may have been some leaching of Ca^{2+} further into the soil profile which was responsible for the marked effect of liming even in season 9 (Table 9.2). Following sunflower, the greater soil N, C, Ca^{2+} and Mg^{2+} were probably responsible for the positive yield responses to the rotation with sunflower. The positive effects of liming and sunflower rotation on infiltration (see Table 9.4) would have further enhanced annual crop yield due to a greater proportion of rainfall being captured in the soil profile.

As elsewhere, land use decisions in vegetable farming systems are driven by farm profitability and economics, and even where good markets for tree products exist, relying on trees alone for a livelihood is not practical, at least for the first few years after planting. Because annual crops offer an immediate return whereas trees are a long-term investment, intercropping of trees and annual crops would appear to be an appropriate strategy from a farm economics perspective. Therefore, integration of trees and annual crops in steepland vegetable production systems could be a promising strategy to minimize soil erosion while ensuring fairly stable farm income. An analysis of the economic trade-offs of intercropping timber and food crops by Nissen *et al.* (2001) showed that even young trees could compete strongly enough for resources to make intercrops unprofitable, and that the length of intercropping in average density stands (approx. 1000 trees/ha) should not exceed one year. Since income from trees may lag by another 2 to 4 years, with the sale of prunings the only interim source of income, farmers are likely to stagger the introduction of trees into their vegetable plots so that they always have vegetables growing among trees of minimal stand basal area. Planting of shade-tolerant crops could, perhaps, extend the intercrop duration without yield penalty. The tree species trialled in this study (*E. grandis*) would appear to be more competitive with food crops than those tested in earlier research (Nissen and Midmore, 2002), although differing pruning practices and planting on land of different slopes, both influencing the effectiveness of the tree canopy in shading food crops, may have differentially influenced competitiveness of tree species. Optimal pruning practices could, therefore, extend the period in which intercropping with annual crops is profitable. Data from our previous work suggest that thinning to one half of plant population (from 5100 to 2550 plants/ha) 12 months after planting favoured stand basal area at harvest more than did maintaining a full population of heavily pruned (removal of 90%

of foliage at 12 months after planting) trees. We do not have comparable data on the net returns of the two contrasting practices. Nevertheless, a call is still made to screen tree species for their suitability for intercropping with food crops during their establishment, to extend the period of income generation from the annual crops in the association.

Conclusions

Identifying appropriate practices for soil erosion control that improve farm income for steep-land vegetable farms is important for the sustainability of commercial vegetable productions systems in the steep-lands of the Philippines and elsewhere. Simply containing erosion by improving soil quality will not lead to adoption of erosion control practices (Hellin, 2004). In our experiments, tree–annual crop intercropping did not minimize tillage-induced and water-related soil loss until natural vegetation was allowed to grow once cropping was curtailed. Planting of annual crops was in rows orientated up and down the slope and removal of prunings from the field was most likely responsible for this result. In-field placement of prunings on the contour would minimize soil erosion due to the containment effect (Stark *et al.*, 2002); clearly the tree canopy alone was not effective in minimizing erosion during the first 2 years of tree establishment. However, developing profitable perennial-based systems is a good strategy for reducing erosion over longer time periods, when the many years of near-zero erosion under perennials are weighed against the more erosive establishment phase.

Growing trees and annual crops together would appear to be more lucrative compared to trees alone, as the DBH in the tree–annual crop system was greater than that of trees alone, and labour costs associated with managing young trees were absorbed by the intercrop. This pattern has important implications for related efforts, such as promotion of cropping systems for carbon sequestration (Shively *et al.*, 2004; Chapter 11, this volume), and the analysis of allocation of resources and land to alternative farming options at a landscape scale (see Chapter 10, this volume). As long as economic incentives exist for the cultivation of high-value vegetable and maize crops in tropical highlands, the promotion of intercropping between trees and annual crops and the development of better land management practices under perennial/annual crop cropping systems will be essential to enhance the long-term sustainability of such farming systems in the Philippines and elsewhere.

Acknowledgements

We acknowledge the SANREM-CRSP funded by the USAID under Grant No. LAG-4198-A-00-2017-00 for supporting this research study. We thank our field staff member, Ferdinand Banda, for data collection, and Hongjun Chen, a postdoctoral research associate at the University of Louisiana-Lafayette, for statistical assistance.

Notes

[1] The photograph in Plate 11 illustrates the typical agricultural landscape in the watershed. A typical vegetable plot is shown in Plate 12.
[2] The estimated relationship is $y = 86.5-70.8x$, $r^2 = 0.91$, where y = % yield of sole cropped annual and x = stand basal area of the trees.

References

Alegre, J.C. and Rao, M.R. (1996) Soil and water conservation by contour hedging in the humid tropics of Peru. *Agriculture Ecosystems and the Environment* 57, 17–25.

Black, C.K. (ed.) (1965) *Methods of Soil Analyses*. Part 2. *Agronomy Monograph* 9. Agronomy Society of America, Madison, Wisconsin.

Ciesiolka, C.A.A., Coughlan, K.J., Rose, C.W. and Smith, G.D. (1995) Erosion and hydrology of steeplands under commercial pineapple production. *Soil Technology* 8, 243–258.

Comia, R.A., Paningbatan, E.P. and Håkansson, I. (1994) Erosion and crop yield response to soil conditions under alley cropping systems in the Philippines. *Soil and Tillage Research* 31, 249–261.

Cramb, R.A. Garcia, J.N.M. Gerrits, R.V. and Saguiguit, G.C. (1999) Smallholder adoption of soil conservation technologies: evidence from upland projects in the Philippines. *Land Degradation and Development* 10, 405–423.

El-Swaify, S.A. (1997) Factors affecting soil erosion hazards and conservation needs for tropical steeplands. *Soil Technology* 11, 3–16.

Gachene, C.K.K., Jarvis, N.J., Linner, H. and Mbuvi, J.P. (1997) Soil erosion effects on soil properties in a highland area of central Kenya. *Soil Science Society of America Journal* 61, 559–564.

Hashim, G.M., Ciesiolka, C.A.A., Yusoff, W.A., Nafis, A.W., Mispan, M.R., Rose, C.W. and Coughlan, K.J. (1995) Soil erosion processes in sloping land on the east coast of Peninsular Malaysia. *Soil Technology* 8, 215–233.

Hellin, J. (2004) From soil erosion to soil quality. *LEISA Magazine* 19, 10–11.

Heusch, B. (1993) Soil erosion in catchments and experimental plots on Java (Indonesia) and Luzon (the Philippines). *Soil Technology* 6, 191–202.

Midmore, D.J. (1993) Agronomic modification of resource use and intercrop productivity. *Field Crops Research* 34, 357–380.

Murphy, J. and Riley, H.P. (1962) A modified single solution method for the determination of phosphate in natural waters. *Analytica Chimica Acta* 27, 31–36.

Nautiyal, S., Maikhuri, R.K., Semwal, R.L., Rao, K.S. and Saxena, K.G. (1998) Agroforestry systems in the rural landscape – a case study in Garhwal Himalaya, India. *Agroforestry Systems* 41, 151–165.

Nelson, D.W. and Sommers, L.E. (1982) Total carbon, organic carbon, and organic matter. In: Page, A.L. *et al.* (ed.) *Methods of Soil Analyses*. Part 2, 2nd edn. *Agronomy Monograph* 9. ASA and SSSA, Madison, Wisconsin, pp. 581–594.

Nissen, T.M. and Midmore, D.J. (2002) Stand basal area as an index of competitiveness in timber intercropping. *Agroforestry Systems* 54, 51–60.

Nissen, T.M., Midmore, D.J. and Keeler, A.G. (2001) Biophysical and economic trade-offs of intercropping *Paraserianthes falcataria* with food crops in the Philippine uplands. *Agricultural Systems* 67, 49–69.

Paningbatan, E.P., Ciesiolka, C.A., Coughlan, K.J. and Rose, C.W. (1995) Alley cropping for managing soil erosion of hilly lands in the Philippines. *Soil Technology* 8, 193–204.

Poudel, D.D., Midmore, D.J. and Hargrove, W.L. (1998) An analysis of commercial vegetable farms in relation to sustainability in the uplands of Southeast Asia. *Agricultural Systems* 58, 107–128.

Poudel, D.D., Midmore, D.J. and West, L.T. (1999) Erosion and productivity of vegetable systems on sloping volcanic ash-derived Philippine soils. *Soil Science Society of America Journal* 63, 1366–1376.

Poudel, D.D., Midmore, D.J. and West, L.T. (2000) Farmer participatory research to minimize soil erosion on steepland vegetable systems in the Philippines. *Agriculture, Ecosystems and the Environment* 79, 113–127.

Presbitero, A.L., Escalante, M.C., Rose, C.W., Coughlan, K.J. and Ciesiolka, C.A. (1995) Erodibility evaluation and the effect of land management practices on soil erosion from steep slopes in Leyte, the Philippines. *Soil Technology* 8, 205–213.

Roose, E. and Ndayizigiye, F. (1997) Agroforestry, water and soil fertility management to fight erosion in tropical mountains of Rwanda. *Soil Technology* 11, 109–119.

SAS (2000) SAS/STAT *User's Guide*. Version 8, SAS Institute, Cary, North Carolina.

Schroth, G. and Lehmann, J. (2003) Nutrient capture. In: Schroth, G. and Sinclair, F.L.

(eds) *Trees, Crops and Soil Fertility: Concepts and Research Methods*. CAB International, Wallingford, UK, pp. 167–179.

Sen, K.K., Rao, K.S. and Saxena, K.G. (1997) Soil erosion due to settled upland farming in the Himalaya: a case study in Pranmati watershed. *International Journal of Sustainable Development and World Ecology* 4, 65–74.

Shively, G.E. (1997) Consumption risk, farm characteristics, and soil conservation adoption among low-income farmers in the Philippines. *Agricultural Economics* 17, 165–177.

Shively, G.E. (2001) Poverty, consumption risk, and soil conservation. *Journal of Development Economics* 65, 267–290.

Shively, G.E., Zelek, C.A., Midmore, D.J. and Nissen, T.M. (2004) Carbon sequestration in a tropical landscape: an economic model to measure its incremental cost. *Agroforestry Systems* 60, 189–197.

Stark, M. Mercado, A., Jr. and Garrity, D. (2002) Natural vegetative strips: farmers' invention gains popularity. *Agroforestry Today* 12, 32–35.

Tacio, H.D. (1993) Sloping agricultural land technology (SALT): a sustainable agroforestry scheme for the uplands. *Agroforestry Systems* 22, 145–152.

10 Linking Economic Policies and Environmental Outcomes at a Watershed Scale*

G. Shively[1] and C. Zelek[2]

[1]Department of Agricultural Economics, Purdue University, 403 West State Street, West Lafayette, IN 47907, USA; e-mail: shivelyg@purdue.edu; [2]United States Department of Agriculture, Natural Resources Conservation Service, 14th and Independence Ave., SW, Washington, DC 20250, USA; e-mail: chuck.zelek@usda.gov

Introduction

Rates of land degradation in most tropical watersheds are greatly influenced by the decisions made by upland farmers. The decisions of smallholder farmers, in turn, are strongly influenced by the interplay of wages, prices and economic opportunities in the general economy (Coxhead, 1997; Shively, 1998; Coxhead *et al.*, 2002; Chapters 4 and 6, this volume). To varying degrees circumscribed by resource constraints, personal goals and societal norms, farmers respond in predictable ways to policy changes instituted at local and national levels. But the ongoing process of economic decentralization and political devolution places pressure on local and national decision makers to define and articulate their spheres of influence and responsibility. This evolution in policymaking raises questions about the appropriate point of entry for policymakers.

In this chapter we use results from a simulation model to study and compare some stylized policy options available to local and national policymakers for whom watershed protection is a goal. Our empirical focus remains on the Manupali River watershed. Our results are based on a modelling strategy that captures, in a very basic way, the interrelationships of biological and economic phenomena. We investigate both the budgetary and human welfare implications of changes in land use induced by a range of policy instruments.

Our approach makes a unique methodological contribution to the field of watershed modelling. The standard approach involves detailed descriptions of

*Parts of this chapter previously appeared in Shively and Zelek (2002) and are reproduced here with permission of the *Philippine Journal of Development*.

©CAB International 2005. *Land Use Change in Tropical Watersheds: Evidence, Causes and Remedies* (eds I. Coxhead and G. Shively)

hydrological and biophysical features, often at the expense of a more life-like characterization of the behaviour of economic agents. In contrast, we present here a model that makes the farm household the basic unit of analysis, and presents household decisions in a way that is consistent with economic theory. We assume optimizing behaviour on the part of farmers, and rational behavioural responses to policy changes. We model environmental externalities as outcomes of optimizing behaviour, and provide indicators of the economic and environmental outcomes associated with land use patterns and policy changes.

The chapter begins with a description of the model structure, then presents the results of four experiments covering a range of prototypical national and local policy changes. Outcomes are measured in terms both of changes in economic welfare and environmental indicators, as well as the implicit costs to implementing agencies. Following this, we discuss the practical and methodological implications of our work, and note some possible extensions of this approach, before presenting a brief conclusion.

Framework, Data and Study Site

The model presented in this section relates economic incentives to decisions and outcomes on representative farms in a heterogeneous landscape. This section provides a brief overview of the model. Additional details can be found in Shively (2002).[1] The mathematical structure of the model can be found in the appendix to this chapter. Figure 10.1 illustrates the process of model construction and utilization of empirical data for calibration and parameterization of the model. As the figure indicates, the model makes use of two main sources of information. Socio-economic data, including panel data described in Coxhead *et al.* (2002), are used to define resources and resource constraints and to develop stylized household models of behavioural response. In addition, land use data are used to calibrate and develop baseline projections of land use and the economic and environmental outcomes of policy changes. Land use data from the watershed also are used to develop weights and scaling parameters that allow us to aggregate outcomes from representative farms and zones up to landscape-scale effects. In developing predictions of erosion outcomes and aggregate impacts, we rely on insights provided by several sources of agronomic and test plot data, including data from the watershed provided by Poudel *et al.* (2000).

In setting up the model we focus on four representative households occupying four representative agroeconomic zones in the Manupali River watershed. As discussed in Chapter 1, the Manupali River watershed is located in north-central Mindanao, in the province of Bukidnon. It extends from Mt Kitanglad in the north-west to the Pulangi River in the south-east. Average elevation is 600 m above sea level and average rainfall is 2300 mm. The watershed covers an area of approximately 38,000 ha, more than 40% of which is classified as hilly or mountainous (Coxhead and Buenavista, 2001). The watershed may be classified into four geomorphic units: mountains, upper footslopes, lower footslopes and alluvial terraces. The population of the

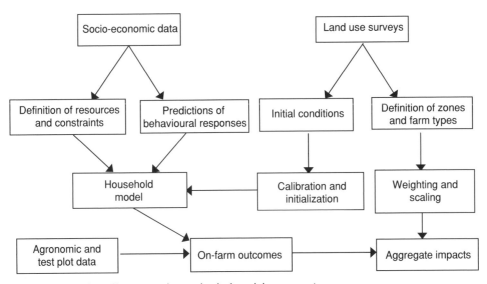

Fig. 10.1. Flow chart illustrating the method of model construction.

watershed in 1994 was 39,500, and until about 2000 the population had grown at an average annual rate of 4% for several decades. In 1988, 71% of employment was in agriculture, 5% in industry, and 23% in services. Major crops grown are maize, sugarcane and rice in the lower elevations, and maize, coffee and a range of vegetables in the upper elevations. Soils in the area are clay and clay loams that are slightly to strongly acidic (see Paningbatan, Chapter 8, this volume). Bin (1994) also describes the physical characteristics of the watershed in detail.

Table 10.1 summarizes cropping patterns for each type of household in the four zones of interest. These zones differ by size, elevation, slope and patterns of cultivation, and broadly characterize four main production regions of the watershed. The representative farms differ according to resource endowments and behavioural characteristics, ranging from farms that are relatively resource constrained and oriented towards subsistence maize production to farms that use purchased inputs (including labour, fertilizer and pesticide) intensively and direct their activity towards commercial vegetable production. These typologies and the data in Table 10.1 are based on detailed farm survey data reported by Coxhead (1995), Coxhead and Rola (1998) and Rola *et al.* (1999). Land use on each representative farm in Table 10.1 is defined in terms of a four-crop portfolio. The table lists primary crops for each household. This typology is a simplification, since in reality many households have the option of choosing from a much wider set of crops. Similarly, few households are strictly limited to a four-crop portfolio. Nevertheless, data in Table 10.1 reflect essential patterns of production in the upper portion of the Manupali River watershed and capture a significant degree of the known variation in land use and outcomes. Most farms grow a combination of food and cash crops; few

Table 10.1. Cropping patterns for representative zones of the watershed.

	Zone 1 (forest-buffer) (%)	Zone 2 (mid-watershed) (%)	Zone 3 (mid-watershed) (%)	Zone 4 (mid-watershed) (%)
White maize	20	60	40	40
Yellow maize	–	–	–	60
Vegetables	60	–	–	–
Coffee	20	40	60	–

Note: Table entries reflect outcomes in base run of the model, computed and calibrated by the authors using data reported in Rola et al. (1999). Land uses in the lower watershed (primarily rice, yellow maize and sugarcane) are not included in the model.

specialize. However, planting patterns suggest that some farmers have better access than others to information and resources such as credit, inputs, hired labour or improved planting materials; these and other factors are likely to determine choices between mainly commercial crops (coffee, vegetables and yellow maize) and the mainly subsistence crop, white maize.

As mentioned in previous chapters, white maize is the most widely grown crop in the upper Manupali River watershed. Nearly all households in the upper reaches of the watershed grow white maize, and in 1997 this crop occupied approximately half of all cultivated area in the upper watershed. Although a market for white maize exists in the area, households typically grow it for home processing, storage and consumption. Following white maize in importance (in terms of planted area) are coffee (30%), yellow maize (12%) and vegetables (10%). Although vegetables occupy a small share of land area, cabbage, potato and tomato production have grown in popularity in recent years, and have economic and environmental importance out of proportion to the land area they occupy.[2] Due to highly erosive and pesticide-intensive production practices, increases in vegetable production have become a concern in the area. Table 10.2 contains average slopes for each zone and erosion rate estimates (by slope) for major crops grown in the zones. Together with land area data, these estimates provide parameters for farm-level erosion predictions in the model.

Our model is designed to simulate optimizing activity in upland agriculture, and its on-farm and off-farm environmental consequences. The logic of the model is guided by choices regarding land shares for available crops. These choices are influenced by relative prices and by policies such as taxes or subsidies on crops, the relative risks of the crops (measured by a variance–covariance matrix for prices), yields, input costs, access to credit, risk aversion and land quality over time. Table 10.3 provides the input requirements and yields of a 1 ha plot of each type of crop. For simplicity it is assumed that input requirements for each crop are the same across zones and households. It is important to note that for modelling purposes, we assume that cabbage provides three crops per year, white and yellow maize provide two crops, and

Table 10.2. Average slopes and estimated erosion rates used in the model.

	Zones 1 & 2	Zones 2 & 3
Average slope (%)	34.5	10.7
Average erosion rate (t/ha/year)		
Coffee	30.0	18.0
Maize	35.0	25.0
Cabbage	55.0	30.0

Note: Slopes calculated using data reported in Bin (1994), Table 5.2. Erosion rates adapted from David (1984), Table 5, using data reported in Cruz *et al.* (1988) and Poudel *et al.* (2000).

Table 10.3. Input requirements and yields per cropping of 1 ha plots in the model.

Crop	Labour (man-days/ha)	Fertilizer (kg/ha)	Pesticide (l/ha)	Yield/ha (kg)
White maize	100	136	0	1000
Yellow maize	100	136	0	1500
Coffee	150	50	0	527
Cabbage	150	211	3	6541

Table 10.4. Watershed model exogenous parameters.

Parameter	Value
Initial soil stock (mm of depth)	3000
White maize price (pesos/kg)	6.50
Yellow maize price (pesos/kg)	5.66
Coffee price (pesos/kg)	39.00
Cabbage price (pesos/kg)	8.00
Labour cost (pesos/man-day)	65.00
Pesticide cost (pesos/l)	421.50
Fertilizer cost (pesos/kg)	7.00
Labour endowment (man-days/household)	400
Farm size (ha/household)	1.00

coffee one crop. Table 10.4 lists price and endowment data used to characterize farms.

For the purpose of deriving off-site outcomes, we identify the lowland sector as a receptor site where sediment accumulates and flows of nutrients and agricultural chemicals can be monitored. Motivation for studying these externalities comes from a study of lowland farms indicating a range of negative impacts, among them siltation in irrigation systems for rice farms (Singh *et al.*, undated). In the model, the flow of erosion from upland households determines the rate of sediment accumulation (and nutrient and pesticide transport) at the receptor site. For any given set of slope and soil characteristics, erosion rates depend on crop shares and area planted, both of which are decisions made by farmers.

The policy dimension of the model consists of crop-specific taxes and subsidies, crop- and technology-specific incentives, and policies to alter price variability in all or specific crops. Values for these parameters characterize the economic environment in which representative farmers make decisions. The model determines crop shares, levels of input use and levels of household income, erosion rates, downstream effects and public sector budgets. Initial values for model parameters and stocks, as well as yield production functions and erosion rates, were derived, where possible, from data collected in the Manupali River watershed.

Figure 10.2 illustrates the model as a system in which variables are related via flows and feedbacks. Households are endowed with initial levels of land, soil quality, labour and capital. Representative farms make choices regarding the allocation of land and other inputs to specific crops. Cropping decisions result in income realization, along with erosion rates. Soil losses are associated with nitrogen and pesticide loss. These result in off-site pollution and sedimentation at the receptor site downstream. In addition, erosion results in a depletion of the farm's soil stock, which reduces productivity in future periods. This is represented by the feedback illustrated in Fig. 10.2. We assume that taxes and subsidies, as well as policies that influence income variability, influence crop choice by altering relative incentives for specific crops, and as a result, also influence environmental outcomes.

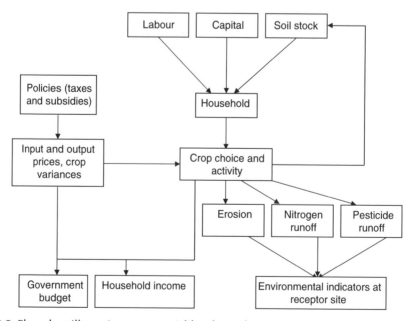

Fig. 10.2. Flow chart illustrating system variable relationships.

Simulation results

We use the model to conduct four policy experiments. These are chosen with the aim of investigating natural resource management policy scenarios that extend across a spectrum, ranging from those emphasizing local focus and control to those that are more broadly based. The policies chosen are: (i) a locally mandated land-use restriction that prohibits vegetable growing; (ii) a locally mandated requirement that farms install and maintain soil conservation structures, combined with a lump-sum subsidy payment for doing so; (iii) a nationally implemented 20% tax on the producer price of vegetables; and (iv) a national/local cost-sharing policy in which revenues from a 20% tax on the producer price of vegetables are used to subsidize the local installation and maintenance of appropriate soil conservation structures. In all cases we conduct our policy experiments over 10-year time horizons and assume the policy changes are sustained for the full length of the simulation. To facilitate comparison, we have converted cash flows to net present values (NPV), using a discount rate of 5%. We express outcomes as indexes relative to outcomes derived in the base run of the model. Results from these simulations are summarized in Table 10.5. For more complete results and discussion, see Shively (2002).

Land use restrictions

Given the importance of vegetable production within the watershed (both in terms of income generation and environmental outcomes) we begin with an investigation of potential economic and environmental impacts of restricting vegetable production. We recognize that a complete ban on vegetable production is neither feasible nor desirable. Nevertheless, simulating such a policy provides insights into outcomes that would be associated with a relatively drastic policy regime. The underlying logic of this policy, which obviously affects only vegetable producers, is as follows. First, the ban on vegetables tends to

Table 10.5. Summary of simulation results.

	Ban on vegetables	Soil conservation	20% vegetable tax	20% vegetable tax + soil conservation
Household income (% change over base)	−15	−0	−12	−15
Govt. budget (surplus or deficit, pesos/ha/year)	0	−391	529	318
Sediment (% change over base)	−37	−46	−15	−49
Nitrogen (% change over base)	−69	−50	−27	−56

Note: Table entries represent cumulative results over a 10-year simulation.

shift land allocation from vegetables into other crops. Under the assumed set of relative prices, this shift is towards coffee production.[3] Second, the ban also encourages some former vegetable farmers to leave a small amount of land in fallow. This pattern is driven by an assumption of an open labour market that allows these farms to sell labour that otherwise would have been used in vegetable production. Under a less favourable off-farm employment scenario, farms would shift labour (and land) into maize production.

The first column of Table 10.5 indicates the impact of the land use restriction. Farm- and area-weighted aggregate income levels are reduced from base levels by approximately 15%. The ban clearly reduces incomes for vegetable growing farms, especially in early years when on-farm erosion has only a minor impact on yields. But the ban reduces incomes only to the extent that alternatives provide lower expected returns than vegetables. Accumulated sediment levels in year 10 are reduced by approximately 37% from base levels. This reflects the shift from vegetables to less erosive forms of land use (coffee and fallow). Nitrogen loadings fall to approximately one-third of base levels, primarily because the nitrogen requirements of vegetables are quite high and are met mainly through the application of inorganic fertilizers. Similarly, since vegetables are the only crops in the model that require pesticides, the vegetable ban leads to an elimination of downstream pesticide loadings. Because the model contains no taxes on land or income, the ban has no direct impact on the government budget.

Soil conservation with a lump sum transfer payment

In the second simulation we investigate the overall impact of requiring the use of soil conservation in the watershed. We implement this policy by assuming labour is required to install and maintain soil conservation structures on each farm, and by deducting this labour availability (25 man-days/ha) from the level of available household labour. At the same time, we subsidize households for this reduction in available labour via an annual transfer payment equal to the value of reallocated labour, where labour is valued at an exogenous and fixed market wage rate. This exogenous wage is lower than the marginal value of labour used in vegetable production. Therefore, from the farm's perspective, the labour used for soil conservation would have been more valuable if used in crop production. For this reason, the technology imposition is not completely income neutral: the reduction in labour availability discourages households from planting vegetables (a labour-intensive crop), and results in an uncompensated income loss. Nevertheless, soil conservation does have the offsetting benefit of increasing yields for all crops, especially in later years of the planning horizon, when yields are most sensitive to the accumulated reduction in the soil stock.

The second column of Table 10.5 shows the impact of this policy. The net present value of aggregate household income remains essentially unchanged over the 10-year period compared with the base run. The policy does induce labour adjustments: some households shift land towards coffee production due to the labour intensity of vegetable production relative to coffee and the loss of labour associated with soil conservation. But for the most part, income losses are offset in the long run by the transfer payment and the yield-maintaining properties of soil conservation. Without the transfer payments, aggregate incomes would fall by approximately 5%. Accumulated sediment is approximately 46% lower in this case than the base case; nitrogen loadings decline by 50%, and pesticide loadings by 52%. The NPV of public expenditures is approximately 391 pesos/ha per year.

It is important to note, however, that the foregoing results are sensitive to assumptions regarding yield–soil loss relationships. The overall change in crop yields over time depends on the shape of the yield functions (with respect to the soil stock) and the assumed rates of erosion. With less responsive yield functions or lower rates of soil loss, the soil conservation policy could result in lower household income due to losses in labour availability. Furthermore, we use a relatively low discount rate of 5%. At higher discount rates, future changes in yields and incomes would count for less, and therefore the soil conservation policy would be less attractive in NPV terms.

A tax on vegetable producers

Results from our third policy experiment, a 20% tax on vegetable producers, suggest a slight increase in the incidence of coffee production and fallow in households that produce vegetables. Results are shown in the third column of Table 10.5. In general, vegetable growers maintain their emphasis on vegetables and sustain a loss in income because the 20% tax is insufficient to generate large changes in land use. Overall, the NPV of income, weighted and aggregated across all household types, falls by approximately 12%. The overall reduction in income is less than the amount of the tax for several reasons: households that do not grow vegetables in the base run of the model see no loss in income, and among vegetable producers, some tax avoidance occurs as households shift away from vegetables. On a watershed basis, the NPV of public revenue with the tax is approximately 529 pesos/ha per year. This average, however, masks some important variability in tax revenue, which fluctuates and ranges from 1800 pesos/ha in year 1 to 47 pesos/ha in year 10. The policy reduces sedimentation by 15% from base levels. Compared with the base, nitrogen and pesticide loadings are reduced by approximately 27% and 39%, respectively, due to the decrease in vegetable production. For vegetable-producing households, some gains in income (*vis-à-vis* the base case) are registered in later years due to decreased cumulative erosion and the beneficial effect this has on yields.

It is important to point out that the main mechanism by which a tax influences an agricultural producer is by changing the relative price he or she receives for a crop – and therefore the expected income from growing the crop. However, the tax also influences the risk–return relationship associated with crops in the portfolio. This second factor is important in the model because some farmers are assumed to be risk averse, that is, concerned about income variability arising from price risk. For this reason, they react to the tax with some degree of friction: they do not fully disengage from production of the taxed crop due to a desire to balance portfolio risks. In other words, the tax decreases household income, but the decrease in the household objective function (which incorporates risk considerations) is less in percentage terms than the tax itself.

Soil conservation in conjunction with a 20% vegetable tax

The final policy we consider combines local incentives to promote soil conservation with what we consider to be a more broadly based (i.e. nationally determined) pricing policy for vegetables. Here we impose the 20% tax on the farm-gate price of vegetables and use the revenues to provide subsidy payments for the use of soil conservation on all farms. Consistent with the tax policy outlined above, we observe increases in area under coffee and fallow. However, farms that produce vegetables have greater incomes relative to the base in some years due, in part, to the yield maintenance afforded by soil conservation.

The NPV of total farm income in the watershed falls by approximately 15% under this policy. Somewhat counter to our initial intuition, income falls by more in this case than under the tax policy alone. The reason for this is that the use of soil conservation causes the household labour constraint to bind. This, combined with the tax on vegetables, discourages their production by more than the tax alone. Yield maintenance is insufficient to compensate. Although detrimental to household incomes, this policy provides strong environmental benefits. Nitrogen and pesticide levels decline by 56% and 62% respectively, due largely to the shift out of vegetables. Sedimentation is likewise reduced by 49%. The government budget shows a net gain, as tax revenues more than offset the subsidy payments. However, the budget does not remain constant over time and the remaining surplus is insufficient to completely offset the losses in income induced by the policy. In years when rates of vegetable production are high the budget is in surplus. In contrast, several years show budget deficit. The government budget ranges from 1231 pesos/ha to −328 pesos/ha, with an average of 318 pesos/ha per year.

Discussion and Potential Extensions

Our results indicate that local bans on crops are expensive (to households) and less effective in curbing erosion than either soil conservation measures, or crop-specific taxes in conjunction with such measures. Furthermore, we find that while income-neutral policies to encourage soil conservation may be useful in reaching erosion targets, they are costly to administer. One alternative approach might be a national/local cost-sharing plan, whereby revenues raised through crop-specific taxes could be used to subsidize local soil conservation initiatives. But due to distortions induced in labour allocation decisions, combining such taxes with local mandates for on-farm soil conservation may be tricky, and could emerge as more costly from the perspective of overall household welfare than tax policies alone.

Several other aspects of our analysis should be noted. First, it may be important to consider the administrative costs of taxation, subsidies and bans. Generally speaking, taxes and subsidies at the farm level are very difficult to administer. This implies that deficits reported in Table 10.5 may be larger and surpluses smaller. Second, any nationally imposed producer tax on vegetables would have a trade impact. A production tax effectively shifts the domestic supply of vegetables to the left, thereby raising imports. This may mean an increase in tariff revenue (in the presence of an import tariff). The Philippines, for example, is currently a net importer of vegetables, including potatoes and tomatoes, suggesting a positive impact on tariff revenue from a production tax. However, a nationally imposed producer tax on vegetables (in general) would also have significant welfare effects on all vegetable farmers, including those in the lowlands (such as onion and garlic farmers). The current model does not incorporate the decisions of lowland farmers, but these could be important in the context of nationally imposed policies like producer taxes.

Finally, it is important to point out that another feasible policy option might be to boost research and extension efforts for high value and environmentally friendly upland crops. For example, yields of upland coffee can be increased through the use of disease-resistant and high-yielding coffee varieties and the Philippine uplands are well suited to production of *Arabica* coffee. But a current problem in the Philippines is that virtually no extension work takes place on *Arabica* coffee, nor on a number of other crops well suited for the uplands.

Conclusion

In this chapter we reported results from an optimization-simulation model used to illustrate potential impacts of local and national policies to encourage sustainable land use. The model was based on a set of four representative households occupying four distinct agroecological zones. Households in the model were assumed to choose crop shares, defined over a portfolio of subsistence food crops, annual and perennial cash crops, to maximize a mean-variance utility function. Using empirically derived population weights, zone-specific transfer coefficients and crop-specific erosion rates, outcomes were aggregated across four household types and four zones to predict watershed-scale changes in land use, incomes and environmental outcomes. The model incorporated risk considerations, labour and land constraints, and labour market opportunities. It provided a stylized view of how households might respond to a range of economic policy changes.

Restrictions on vegetable growing reduced farm- and area-weighted income levels by approximately 15% over 10 years, and reduced downstream loads by up to 37% from base levels. Soil conservation, combined with a lump sum transfer payment, was somewhat more effective in reducing erosion and associated agricultural externalities, but at a cost to the government of approximately 391 pesos/ha per year. Simulation results suggest a 20% tax on vegetable production is insufficient to dramatically alter land use patterns. Households shift land use somewhat to avoid taxes, leading to modest reductions in agricultural externalities. Combining the tax with a subsidy to soil conservation provides strong or environmental benefits, but reduces household welfare. In general, we find that economy–environment trade-offs are difficult to avoid. Policies that generate large environmental improvements (measured in terms of reduced sediment and reduced nitrogen and pesticide flows) tend either to reduce household incomes in the short or long run, or to place budgetary burdens on local government. However, our analysis demonstrates the potential value of conducting analysis of economy–environment interactions at a landscape scale of analysis. Better policy coordination between national and local policymakers, and between those responsible for agricultural, trade and environmental policies, appears to be a necessary prerequisite for achieving least-cost environmental protection.

Notes

[1] Also see http://www.agecon.purdue.edu/staff/shively/manupali
[2] Cabbage is locally important in terms of value. In terms of its production and market characteristics cabbage is representative of a range of vegetable crops grown in the area, including potato and tomato.
[3] This substitution of coffee for vegetables in the model reflects a favourable relative price for coffee. With a lower relative price for coffee, farms would instead reallocate land to maize production.

References

Bin, L. (1994) The impact assessment of land use change in the watershed area using remote sensing and GIS: a case-study of Manupali watershed, the Philippines. MS thesis, School of Environment, Resources, and Development, Bangkok.

Coxhead, I. (1995) The agricultural economy of Lantapan, Bukidnon, Philippines: results of a baseline survey. Social Science Working Paper No. 95/1. SANREM CRSP-Philippines.

Coxhead, I. (1997) Induced innovation and land degradation in developing country agriculture. *Australian Journal of Agricultural and Resource Economics* 41, 305–332.

Coxhead, I. and Buenavista, G. (eds) (2001) Seeking Sustainability: Challenges of Agricultural Development and Environmental Management in a Philippine Watershed. PCARRD, Los Baños, Laguna, Philippines, 267 pp.

Coxhead, I. and Rola, A. (1998) Economic development, agricultural growth and environmental management: what are the linkages in Lantapan? Paper prepared for SANREM CRSP-Philippines conference, Malaybalay, Bukidnon, Philippines, 18–21 May 1998.

Coxhead, I., Shively, G. and Shuai, X. (2002) Development policies, resource constraints, and agricultural expansion on the

Philippine land frontier. *Environment and Development Economics* 7, 341–364.

Cruz, R., Francisco, H. and Conway, Z. (1988) On-site and downstream costs of soil erosion in the Magat and Pantabangan watersheds. *Journal of Philippine Development* 15, 85–112.

David, W.P. (1984) Environmental effects of watershed modifications. Philippine Institute for Development Studies Working Paper 84–07. Philippine Institute for Development Studies, Manila.

Poudel, D.D., Midmore, D.J. and West, L.T. (2000) Farmer participatory research to minimize soil erosion on stee-pland vegetable systems in the Philippines. *Agriculture, Ecosystems, and the Environment* 79, 113–127.

Rola, A., Tabien, C. and Bagares, I. (1999) Coping with El Niño, 1998: an investigation in the upland community of Lantapan, Bukidnon, Philippines. Working Paper 99–03. Institute of Strategic Planning and Policy Studies, University of the Philippines at Los Baños, Philippines.

Shively, G. (1998) Economic policies and the environment: the case of tree planting on low-income farms in the Philippines. *Environment and Development Economics* 3, 15–27.

Shively, G. (2002) Conducting Economic Policy Analysis at a Landscape Scale: An Optimization-simulation Approach with Examples from the Agricultural Economy of a Philippine Watershed. Research monograph, Purdue University. Accessed at www.agecon.purdue.edu/staff/shively/manupali

Shively, G. and Zelek, C. (2002) Linking economic policies and environmental outcomes at a watershed scale. *Philippine Journal of Development* 29(1), 101–125.

Singh, V.P., Bhuiyan, S.I., Price, L. and Kam, S.P. (Undated) Development of sustainable production systems for different landscape positions in the Manupali River watershed, Bukidnon, Philippines: soil and water resource management and conservation. SANREM CRSP Phase I Terminal Report.

Appendix

The optimization-simulation model used for this analysis was constructed using Excel and STELLA. The latter is a computer simulation software package designed to model dynamic systems. The model consists of four representative upland farms that make crop portfolio decisions under uncertainty. We use a mean-variance framework for the household model in which a farmer attempts to maximize expected utility (assumed to be a weighted-sum of mean returns and variance in returns), subject to a set of resource (land and labour) constraints. The decision maker is assumed to optimize his farm plan at the start of each year, but we assume that the farmer is myopic with respect to the future impacts of current decisions. We define variables in the model as follows: θ_i is the share of land planted with crop i, β_i the mean return for crop i, σ_{ih} the price variance for $i = h$ and the covariance for $i \neq h$, and ρ the coefficient of risk aversion. Risk sensitivity in the model is controlled via this farm-specific risk aversion parameter. The mean return for each crop is computed as the market price for the respective crop p_i adjusted for the tax imposed on that crop t_i, multiplied by the crop-specific yield y_i, after which input costs, c_i, net of any subsidies s_i, are deducted. For simplicity we assume a unit cost function for each crop and normalize planted area on the farm to be 1 ha. Each household has a set of choice variables $\{\theta_i\}$ at each year in the simulation. The household level problem is:

$$\max_{\theta_i} \sum_{i=1}^{n} \beta_i \theta_i - \frac{\rho}{2} \sum_{i=1}^{n} \sum_{h=1}^{n} \sigma_{ih} \theta_i \theta_h \tag{10.1a}$$

$$s.t \sum_{i=1}^{n} \theta_i \le 1 \tag{10.1b}$$

$$\theta_i \ge 0 \tag{10.1c}$$

where

$$\beta_i = (1 - t_i) p_i y_i - (1 - s_i) c_i \tag{10.1d}$$

To incorporate policy parameters that might possibly affect income variance, the variance and covariance terms are modified as follows:

$$\sigma_i^* = \sigma_i (1 + \gamma_i) \tag{10.2a}$$

and

$$\sigma_{ih}^* = \sigma_{ih} \sqrt{(1 + \gamma_i)(1 + \gamma_h)} \tag{10.2b}$$

where γ_i is the variance-adjusting policy parameter for crop i. Examples of policies that may reduce variance are those that reinforce the marketing infrastructure for agricultural crops, for example, the transportation system through road construction. It is possible to assess the impact of crop-specific policies designed to reduce income variability for targeted crops. Examples of policies that could reduce income variability for specific crops include research into pest-resistant varieties or programmes targeted at improving post-harvest handling for certain crops. Furthermore, policies that reduce variability of certain crops could likewise increase mean values for the respective crops. However, this effect is not included in this model due to data limitations.

Erosion is measured at the farm level. Erosion depends on both physical phenomena and crop composition. We posit the farm-level function for erosion E:

$$E_{k,t} = f(\theta_i, G, T, R) \tag{10.3}$$

where erosion on farm k at time t is a function of crop composition, slope gradient G, soil type T and rainfall R. We expect the slope gradient to be positively related to erosion and for the erosion rate to increase at an increasing rate as slope steepness increases. A similar relationship holds for rainfall: the soil's ability to absorb rainfall decreases as the amount of rainfall increases, thus increasing the rate of erosion. Erosion is measured as a flow. In each period erosion decreases the stock of soil available on the farm via the equation of motion:

$$S_{k,t} = S_{k,t-1} - E_{k,t} \qquad (10.4)$$

In other words, the soil stock at any time t equals the soil stock at time $t-1$ minus the flow of erosion in period t. As noted above, this is a farm-level outcome. Erosion also increases the stock of sediment accumulating off-site. The stock of sediment Q_t accumulates over both time and space. Defining w_k as a farm-specific erosion-sediment transfer coefficient and δ_j as a zone-specific sediment-delivery delay parameter, accumulation of sediment at the receptor site is:

$$Q_t = Q_{t-1} + \sum_{\tau=0}^{t} \sum_{j=1}^{m} \delta_j \sum_{k=1}^{q} w_k E_{k,\tau} \qquad (10.5)$$

where t equals the number of periods in the history of the simulation, m equals the number of zones, q equals the number of representative farms within each zone, and E is defined as above. Note that past erosion events contribute to sediment according to a delay specified by δ_j. Site-specific erosion affects sediment via a conveyor process: the stock of sediment at time t is equal to the stock at time $t-1$ plus the flow of erosion from zone j that reaches the receptor site by t.

Viewed from the perspective of erosion and its local affect on the soil stock, the model described thus far characterizes a simple dynamic model with a positive feedback loop. During each planting season the farmer chooses an optimal crop portfolio. This generates erosion. The flow of erosion decreases the soil stock and reduces subsequent yields. We model this feedback explicitly by formulating crop-specific production functions of the form:

$$y_{i,k,t} = f_i\left(K_{i,k,t}, L_{i,k,t}, S_{k,t}\right) \qquad (10.6)$$

The model allows for the measurement of welfare at the level of the household, the zone or the watershed. These welfare measures can be computed at a point in time, or can be expressed as the present discounted values of the stream of incomes, summed across households. Using r to represent the discount rate and $\beta_{k,t}$ to represent net income at time t for household k (i.e. $\Sigma\theta_{i,t}\beta_{i,t}$), and using T and v to represent area weights for representative households and zones, respectively, the NPV formula for the household is:

$$W_k = \sum_{t=0}^{T} \left(\frac{1}{1+r}\right)^t \beta_{k,t} \qquad (10.7)$$

For the zone it is:

$$W_j = \sum_{k=1}^{q} \omega_k \sum_{t=0}^{T} \left(\frac{1}{1+r}\right)^t \beta_{k,t} = \sum_{k=1}^{q} \omega_k W_k \qquad (10.8)$$

And for the watershed it is:

$$W = \sum_{j=1}^{m} \upsilon_j \sum_{k=1}^{q} \omega_k \sum_{t=0}^{T} \left(\frac{1}{1+r}\right)^t \beta_{k,t} = \sum_{j=1}^{m} \upsilon_j W_j \qquad (10.9)$$

A partial measure of the impact of policies on the government budget can be calculated based on taxes and subsidies applied to crops or inputs. The budget is partial in the sense that we can compute the costs of taxes and subsidies on crops or inputs, but cannot compute the costs of policies that might effectively stabilize markets (e.g. infrastructure improvements) or alter land tenure arrangements.

In terms of the entire watershed, the budget impact of a policy depends on the intervention (tax or subsidy rates), the behavioural outcomes at the farm level (e.g. crop choice), and the household or zone-specific weights applied to representative households. The NPV computation is:

$$B = \sum_{t=0}^{T} \left(\frac{1}{1+r}\right)^t \sum_{j=1}^{m} \upsilon_j \sum_{k=1}^{q} \omega_k \sum_{i=1}^{n} \theta_{i,k,t} \left(t_{i,t} p_{i,t} y_{i,k,t} - s_{i,t} c_{i,t}\right) \qquad (10.10)$$

where variables are defined as above. Time subscripts in this equation indicate tax or subsidy rates may change over time.

11 Using Payments for Environmental Services (PES) to Assist in Watershed Management*

S. PAGIOLA,[1] M. DELOS ANGELES[2] AND G. SHIVELY[3]

[1]The World Bank, 1818 H Street NW, Washington, DC 20433, USA; e-mail: spagiola@worldbank.org; [2]The World Bank, 1818 H Street NW, Washington, DC 20433, USA; e-mail: mdelosangeles@worldbank.org; [3]Department of Agricultural Economics, Purdue University, 403 West State Street, West Lafayette, IN 47907, USA; e-mail: shivelyg@purdue.edu

Overview

Paying land users to provide environmental services is a potentially useful way to protect key environmental resources in areas where they are threatened by degradation or overuse. The payments-for-environmental-services (PES) approach is based on two main principles: that those who provide valuable environmental services should be compensated for doing so; and that those who benefit from these services should pay for them. In theory, this approach has the potential to greatly increase the efficiency with which environmental services are protected. At the same time, because there is a high degree of spatial correlation between areas of rural poverty and locations of key environmental resources (Nelson and Chomitz, 2002), payments can serve a second purpose by providing additional income for low-income individuals. This aspect of PES systems suggests they may contribute to poverty reduction at the same time as they address environmental concerns (Pagiola et al., 2005).

PES programmes have been implemented in several countries, including Costa Rica, Colombia, Ecuador and Mexico. The basic framework for establishing PES programmes can be found in a number of countries, including the Philippines. In this chapter we briefly review the logic behind PES and examine

*Some of the material in this chapter is drawn from Pagiola and Platais (forthcoming); permission to use this material is gratefully acknowledged. All opinions expressed here are those of the authors and do not necessarily represent those of the World Bank or sponsoring agencies.

the potential for using a PES programme to address environmental and poverty concerns in the Manupali River watershed.

Why Pay for Environmental Services?

Previous chapters in this volume have documented the ongoing conversion of forest into agricultural land in the Manupali River watershed over the past 20 years and the impact of relative price changes in precipitating changes in land use. It is clear from the evidence provided (e.g. Coxhead and Demeke, Chapter 5, this volume) that land users make rational land use choices, driven in large part by the net benefits they expect to obtain from competing land uses. At the same time, the resulting patterns of land use have had adverse impacts, especially downstream. As Deutsch and Orprecio (Chapter 3, this volume) and Ella (Chapter 7, this volume) demonstrate, the watershed has experienced much higher erosion rates than one would expect from an undisturbed landscape, and the instability of some tributaries (indicated by abrupt flooding and drought cycles) appears to be intensifying. Less obviously, adverse impacts from forest loss and degradation have occurred at a much higher level. The global community as a whole is harmed by damage to the region's biodiversity and by reduced carbon sequestration, which contributes to global climate change.

Current agricultural land uses are undoubtedly among the most profitable for farmers, but they produce a much lower level of valuable environmental services such as watershed protection, biodiversity conservation and carbon sequestration than the previous, forested land use or alternative land uses, such as those outlined by Midmore *et al.* (Chapter 9, this volume). The source of this disconnect between private and social valuations is simple: regardless of how valuable off-site environmental services may be to those who benefit from them, they typically provide few benefits to those responsible for making actual land use decisions. Unsurprisingly, these land users tend to ignore the off-site effects when making land use decisions.

Introducing payments for environmental services is one potential way to address this mismatch between the upstream and downstream valuations of activities and resources. Payments by downstream service users can help make land uses that generate environmental services more attractive to upstream land users. As long as these payments exceed the opportunity cost to land users of engaging in degrading activities, and are less than the value of the benefits service users receive, the arrangement is in the interest of both parties. Such a PES programme could be more efficient than a traditional command-and-control approach (Pagiola *et al.*, 2002; Pagiola and Platais, forthcoming). The costs of achieving any given environmental objective are rarely constant across all situations or potential suppliers, even within a watershed, and PES programmes take advantage of these differences by concentrating service effort where costs are lowest. Likewise, the benefits arising from conservation efforts can differ substantially across beneficiaries. A PES programme can concentrate effort where it is likely to provide the greatest benefit. Moreover, by basing payments to service providers on payments from service users, PES programmes have a built-in feedback mechanism: service users have a strong incentive to ensure that their money is spent effectively, and to request changes in the programme if it is not. These characteristics add to the likelihood that a PES programme will be sustainable over time because the system depends on the self-interest of the affected parties, rather than the philanthropy or whims of donors. For the same reasons, there is a presumption that PES programmes will make upstream participants better off (and certainly no worse off) resulting in the likelihood that a PES programme can help to reduce poverty.[1]

Putting PES Theory into Practice

A number of PES programmes have been undertaken in recent years (Landell-Mills and Porras, 2002; Pagiola *et al.*, 2002; Pagiola and Platais, forthcoming). Latin America has been a particularly fertile ground for PES initiatives. Two countries have created nationwide PES programmes. Costa Rica has led the way, with its *Pago por Servicios Ambientales* (PSA) programme, under which land users can receive payments for specified land uses, including new plantations and conservation of natural forests (FONAFIFO, 2000; Pagiola, 2002). In 2003, Mexico created the Payment for Hydrological Environmental Services programme – *Pago por Servicios Ambientales Hidrológicos* (PSAH) – which pays for the conservation of forests in hydrologically critical watersheds using revenue from water charges (Bulas, 2004).

The PES approach has also been used for biodiversity benefits in a few cases. While in general using a PES approach for biodiversity conservation services may be much harder than for either water services or carbon sequestration services (because it is very difficult to identify the beneficiaries of the service[2]), the GEF-financed Regional Integrated Silvopastoral Ecosystem Management Project (RISEMP) is using a PES approach to generate biodiversity conservation services by encouraging the adoption of silvopastoral practices in Colombia, Costa Rica and Nicaragua (Pagiola *et al.*, 2004). Environmental NGO Conservation International has also used the approach (which it calls 'conservation concessions' or 'conservation incentive agreements') in Guyana and Peru (Hardner and Rice, 2002; Rice, 2003).

Despite several nationwide efforts to establish PES systems, the trend has been towards initiatives at the scale of individual watersheds. Many municipal water supply systems have adopted PES approaches to protect their water supplies. They range in size from several million people in Quito, Ecuador, to 3800 people in Yamabal, El Salvador. Irrigation water user groups (e.g. in Colombia's Cauca valley) and hydroelectric power (HEP) producers (in Costa Rica, for example) are also paying to conserve the watersheds that supply them with water.

Although the PES approach is intuitively appealing, putting it into practice is far from simple. A PES programme requires that one can identify and measure benefits and costs associated with the provision of environmental services, and properly attribute the benefits and costs across a landscape. It requires convincing service users to pay for the services, and it requires a mechanism to collect and manage these payments. It also requires a mechanism to contract for land users to adopt the desired land uses, monitor compliance and make payments. Weak institutions, tenure insecurity and pre-existing policy distortions are common stumbling blocks to the implementation of PES programmes.

In general, putting a specific PES into practice in a specific watershed involves four broad, overlapping steps: (i) identifying and quantifying the environmental services involved; (ii) developing financing mechanisms that capture some of the benefits obtained by services users; (iii) developing compensation mechanisms to pay the service providers; and (iv) developing the institutional structure to implement these mechanisms.

Identifying environmental services

Although ecosystems provide a wide variety of services, identifying and valuing the services generated by different land uses is difficult. This is partly because of the diversity and complexity of the conditions encountered and partly because of the diversity of objectives being sought. For example, hydrological benefits may depend on the rainfall regime, on the type of soil and vegetation in an area and on topography.[3]

Developing financing mechanisms

Service users typically do not receive generic 'ecosystem services' but are instead interested in very specific services. Even within specific service categories, differences arise. So, for example, domestic water supply systems may be interested in a constant flow of high-quality water while a hydroelectric power producer may care little about water quality except for the absence of sedimentation. The willingness of a given group of service users to pay will depend on the specific service they receive, on the value of that service to them (compared with the cost of alternatives) and on the size of the group. Once the users of a service are known, a means must be devised to translate their willingness to pay into a stream of payments. This is obviously easiest when the service users are easily identifiable and are already organized, making it relatively simple to negotiate with them and to collect payments. For example, an additional fee can be added to water bills paid by municipal and industrial water users.

Developing effective compensation systems

In a PES system service providers must be compensated for services they provide. A primary challenge is doing so in a way that both generates the desired services and keeps transaction costs low. PES programmes will have the desired effect only if they reach the land users in ways that influence relevant land use decisions. The best systems will be continuous and open-ended, targeted and not responsible for generating perverse incentives (such as encouraging land users to cut down standing trees so as to qualify for reforestation payments).

Developing a supporting institutional structure

Finally, an institutional infrastructure must be established to broadly mirror the three steps outlined above: technical institutions to identify environmental services and the best way to provide them; financial institutions to capture benefits from service users and handle the funds; and field-based institutions to elicit service provision from land users, monitor compliance and make payments. On top of this, a governance structure must be established to bring together the PES programme stakeholders. More broadly, rights to the property providing environmental services must be clearly defined, and the legal framework must, at a minimum, not prevent the establishment of a PES programme. In many cases, parts of the required institutional structure will already be in place, but may be required to undertake new functions.

PES in the Philippines

The opportunity for PES in the Philippines rests on several pieces of national legislation as well as on recent global agreements on carbon reductions. The most important local legislation guiding PES is the National Integrated Protected Area System (NIPAS) Act of 1992, which empowers the Department of Environment and Natural Resources (DENR) to 'fix and prescribe reasonable NIPAS fees to be collected from government agencies or any person, firm or corporation deriving benefits from the protected areas'. Although this language seems to apply to any environmental service, including those which only provide indirect benefits, in practice the law has been applied mainly to direct uses of resources, such as revenues generated from entrance fees or resource extraction (REECS, 2003). In 2003, Mt Kitanglad Range National Park (MKRNP), a portion of which borders the Manupali River watershed, generated 500,000 pesos in entrance and other fees (Rola *et al.*, 2003). Funds generated in this way are placed into a special fund known as the Integrated Protected Area Fund (IPAF). In principle, these funds could be used to fund a PES programme: 75% of the collected fees are supposed to be returned to the protected area (PA) that generated them.[4] In general, NIPAS fees could be used to support PES programmes in a variety of ways.

PES for water

Water users include HEP facilities, downstream agricultural producers and municipalities. In the case of HEP facilities, an existing fee structure provides another potential starting point for a PES programme. Across the Philippines, operators of HEP facilities pay a fee of 0.5 centavo per kilowatt-hour of electricity sales. Roughly one quarter of this amount must be placed into a special account called a Reforestation, Watershed Management, Health and/or Environment Enhancement Fund (RWMHEEF). A portion of the RWMHEEF is earmarked for watershed management, although this has not usually taken the form of PES to date (Winrock International, 2004). Instead, the National Power Corporation (NPC), which operates the Pulangi IV HEP plant downstream from the Manupali River watershed, has undertaken its own watershed management activities in recent years. Reorienting RWMHEEF funds towards a PES-type system may be desirable.

An additional source of PES funds tied to water is the National Irrigation Administration (NIA). NIA currently operates the 4400 ha Manupali River Irrigation System (MRIS) near the city of Valencia, as well as several other irrigation systems in Bukidnon Province. The MRIS has been adversely affected by degradation in the Manupali River watershed: dredging costs in the MRIS have increased substantially (Rola *et al.*, 2004) and rice yields in MRIS farms affected by siltation fell by 27% from 1990 to 1995 (Lantican *et al.*, 2003). Although the MRIS has not actively supported watershed management to date, it spends some 700,000 pesos annually on desilting irrigation canals and repairing structures in its irrigation systems (Rola *et al.*, 2004). Therefore, the MRIS might be willing to contribute to watershed management activities that promised to reduce these costs significantly.

Note that there are some water use issues in the Manupali River watershed that PES is unlikely to be able to address. Two banana plantations, established in 1999, are expected to consume some 22,500 m^3 of water a day, and several commercial livestock operations also use considerable amounts of water (Rola *et al.*, 2004). The high level of water demand from these activities creates potential water use conflicts with users further downstream. They are also potentially significant sources of contamination. It is unlikely that PES mechanisms could easily address these problems. There is also a need to identify other sediment sources, such as poorly designed or constructed roads, that may also be important sources of water degradation.

PES for carbon sequestration

The Clean Development Mechanism (CDM) of the Kyoto Protocol provides a mechanism for rewarding low-income countries for the provision of carbon sequestration services. A number of schemes are currently being developed to take advantage of the CDM and the potential for carbon sequestration in low-income areas.[5] At the moment, reforestation is the only land use-based method of obtaining emission reduction under the CDM. Given the large area of Manupali River watershed that has been deforested, there may be considerable scope to generate Kyoto-compliant emission reductions. The market for non-Kyoto emission reductions, which has less restrictive rules but also much lower prices, is smaller. In 2003–2004, Kyoto-compliant emission reductions sold for US$3.85–5.52/t of CO_2 equivalent (tCO_2e) (volume-weighted averages), depending on whether the buyer or the seller took the risk of non-ratification, while non-Kyoto-compliant emission reductions sold for US$1.34 per tCO_2e (World Bank, 2004).

For the Manupali River watershed Zelek and Shively (2003) examined the costs of sequestering carbon via conversion of annual crop agricultural land to forestry or agroforestry systems, based on the opportunity cost of current land uses. Their analysis shows that, as the total amount of carbon sequestered rises, the opportunity cost of land conversion increases due to both changes in land quality and changes in land use. An agroforestry system was found to be a somewhat lower-cost alternative to pure forest conversion, with average per-ton carbon costs that were approximately 8–16% lower than the costs for carbon storage via a pure tree stand. The estimated cost of sequestering carbon over a 10-year period ranged from US$ 0.90/tCO_2e on fallow land to US$17.0/$tCO_2e$ on land planted to high-value crops.[6]

One potential problem with designing a PES system around smallholder carbon sequestration services is that, from the prospective of global beneficiaries of the environmental service, carbon sequestration is completely generic. One tonne of C sequestered on a low-income farm is equivalent to a tonne sequestered at an alternative source. Although Zelek and Shively's estimates demonstrate that carbon sequestration in the Manupali River watershed could be cost competitive with other emission reduction sources, if transaction costs for sequestering carbon in this way are substantially higher than for other sources of carbon reduction, then they may erase any opportunity cost advantages provided by these suppliers. For this reason, there may be limited scope for benefiting the poor through carbon-based PES schemes, unless they can be 'piggy-backed' on some other activity that is already bearing the brunt of transaction costs, or a third party such as an NGO or donor subsidizes the transaction costs.

Opportunity Costs and Participation Patterns

By design, PES programmes are voluntary. It follows that payments under a PES programme must be sufficient to make land users at least as well off as under their current land use, if they are to be induced to participate. The opportunity cost of participation varies, of course, across potential participants, depending on both the current activity and the quality of resources under consideration. Zelek and Shively's (2003) study provides a way to estimate the opportunity costs faced by farmers in the Manupali River watershed of switching from their current land uses to either a forestry system or an agroforestry system. Figure 11.1 indicates these opportunity costs, measured in present value terms, over 10 years, discounted at 12%. As might be expected, opportunity costs are lower on low-input than on high-input farms. They are also lower, for a given farm, the lower the land quality. The opportunity cost of switching to a forest system is higher than that of switching to agroforestry, due to the lower costs and higher on-farm benefits of the latter, though the difference is minimal on low-input farms.

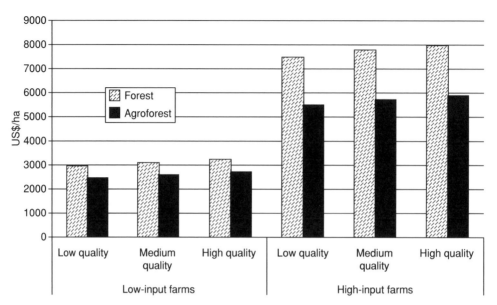

Fig. 11.1. Opportunity cost (in US$/ha) of switching from annual crops to tree-based systems, by land quality and farm type, Manupali River watershed. (*Source*: based on data in Zelek and Shively, 2003.)
Note: Present value over 10 years, discounted at 12%.

What do the data in Fig. 11.1 tell us about the potential role of a PES programme in Manupali? First of all, they suggest something about the minimum payments that would be required to induce different groups of farmers to participate. Broadly speaking, low-input farmers would need total payments of just over US$3000/ha over a 10-year period to alter land uses (although slightly less for conversion to agroforest).[7] High-input farmers, on the other hand, would require payments of at least US$5500–5900/ha to be induced to switch to agroforestry, and of at least US$7400–8000 to be induced to adopt a forestry system.[8]

Such payment schemes might play out as follows. Consider, first, a PES programme that offers a payment worth US$2900/ha for adoption of either forestry or agroforestry systems for an ensuing decade. This PES would attract all low-input farms and no high-input farms, and all participants would choose to undertake agroforestry rather than forestry (as the opportunity cost is lower). Based on data from the watershed, this suggests that at most 22,900 ha would be enrolled and 18,800 ha would not. If poorer farmers tend to have low-input farms, then a large proportion of participants will be among the poorest farmers.

If forestry systems are preferred to agroforestry systems from an environmental service perspective (as would be the case when either biodiversity conservation or carbon sequestration is the main objective), the same US$3000 payment offered for forestry systems (but not for agroforestry) would induce only low-input farms on low-quality land to enroll. Low-input farms on medium- and high-quality land and all high-input farms, in contrast, would decline to participate.

These flat payments would be the simplest to implement, and would provide the greatest area enrolled at the lowest possible cost, so long as the location of enrolled areas does not matter. Sometimes, however, location does matter. Location considerations might require targeting a specific watershed, or targeting a particular area within a watershed. So, for example, if reducing erosion is the principal concern, then based on the results presented by Deutsch and Orprecio (Chapter 3, this volume) one would want a PES programme to target the Kulasihan River watershed. Moreover, within this watershed, the results presented by Paningbatan (Chapter 8, this volume) suggest that the PES programme should ideally target hotspot areas with steep slopes or particularly erosive soils. Conversely, if biodiversity conservation is the main objective, the PES programme might target the upper parts of the watershed, in the buffer zone of the MKRNP.

Poverty Impacts

PES programmes are not intended for poverty reduction, but well-designed programmes conceived under favourable local conditions may reduce poverty. Overall, the fact that participation in a PES is voluntary creates a strong presumption that participants are better off, particularly for service providers who receive payments. If this were not the case, they could simply refuse to participate, or end their participation. The extent to which they are better off is an empirical matter, which has been little studied to date. It clearly depends substantially on the amount of the payment, and on the opportunity costs landowners must bear to take part, including the cost of forgoing alternative land uses, and any transaction costs that participation may entail.

Following the logic of the exercise outlined above, consider a biodiversity-based targeting scheme in the Manupali River watershed. In general, the upper parts of the watershed tend to be of lower-quality land, and farms tend to use lower input levels. A payment of roughly US$2900/ha over 10 years would probably attract the majority of farmers in the target area and would probably impact predominantly poor farmers. Conversely, an erosion-based targeting would probably include areas with high-quality lands and high-input farms. In these areas, the same payment would probably attract only a small proportion of all farmers – possibly too few to result in an improvement in service provision – and would probably reach a proportionately smaller share of poor farmers. Offering higher payments to attract a larger number of participants, say into an agroforestry scheme, would again tend to reduce the share of relatively poor farmers in the overall programme. If the programme targeted forestry systems only, then the necessary payments would be even higher. The cost of obtaining the services would be contained, on the other hand, because only a portion of the total watershed would be eligible to participate.[9] But this portion might possibly include the better-off segment of the farm population.

Finally, even if upstream land users are likely to benefit from PES, there may not automatically be a substantial poverty impact. Unfortunately, many factors that might prevent or limit participation in a PES programme are likely to be correlated with poverty. These include insecure land tenure, lack of title, small farm holdings and lack of access to credit. The extent to which these problems will prove to be obstacles in practice remains to be seen. Much will depend on the specific characteristics of the PES programme and the conditions under which it is implemented. This points to several ways in which PES programmes can be designed to try to minimize adverse impacts and maximize positive ones. Probably the most important step is to design the payment mechanism so as not to exclude poor land users. This requires keeping the transaction costs as low as possible, and being creative in response to problems such as insecure tenure or lack of titles. This will be easier to do when there are strong local organizations such as community groups or NGOs that can help organize participants and provide a forum for discussing solutions to problems as they arise.

In considering how best to design a PES programme so as to improve its poverty impact, it is important not to fall into the trap of considering the programme as being primarily a poverty reduction tool. If the objective of generating services is subordinated to that of poverty reduction, then the PES programme may fail to deliver services, thereby undermining the very basis of the programme. PES programmes *cannot*, for example, target areas of high poverty, as these may not be the areas that generate the desired services. Within an area that generates services, however, they *can* try to design the payment mechanism so as to allow the poor to participate.[10]

Conclusions

PES is not a universal solution to environmental problems. It is well suited to situations in which identifiable land uses produce substantial positive externalities, and where the characteristics of service users and service providers are such that payment programmes have limited transaction costs. For these reasons, expecting PES to solve all of Manupali's problems would not be realistic. There appears to be a clear potential to use both water payments and carbon payments to encourage improved land uses, in at least part of the watershed. Water payments would probably focus on high-erosion parts of the watershed, while carbon payments would focus on low-opportunity parts of the watershed. The degree to which these overlap (and, hence, the potential for bundling payments) remains to be determined. In low-opportunity cost parts of the watershed, carbon payments would probably be sufficient by themselves, if transaction costs can be kept at acceptable levels, to induce fairly substantial land use change. Payments in high-erosion parts of the watershed would probably have to be higher, as they would have to displace land uses with high opportunity costs (such as commercial vegetable production). The potential for biodiversity payments is less clear, as there is no obvious financing source except for IPAF, which at the moment can only be used for projects inside defined protected areas. Implementing either water payments or carbon payments would entail significant transaction costs, as there is currently no structure in place to either contract with land users for specific land uses or receive and handle funds. Moreover, it remains to be determined whether benefits to service beneficiaries would be sufficient to justify environmental payments. Nevertheless, as local institutions continue to strengthen, policymakers might fruitfully consider incorporating PES approaches within their overall mix of strategies to address environmental concerns in the watershed.

Notes

[1] As an example of how such a programme might improve overall well-being, Shively (forthcoming) shows how a transfer payment scheme that redistributes income from lowland farmers to upland farmers in exchange for reduced upland cultivation can, under the right set of circumstances, benefit both groups and reduce relative poverty.

[2] In the early and mid-1990s, biodiversity prospecting ('bioprospecting') was expected to provide an important new source of financing for forest conservation (Principe, 1989; Reid *et al.*, 1993). Much-publicized contracts, such as that between pharmaceutical company Merck and Costa Rica's InBio, increased the expectation that this approach could lead to substantial resource flows for conservation. These hopes have largely not been realized. Moreover, recent studies have shown relatively low values for conservation. Simpson *et al.* (1996), for example, examined pharmaceutical researchers' willingness to pay for biodiversity as a means for financing the protection of higher plant species' biodiversity and genetically diverse habitats. They estimated the value of willingness to pay to preserve genetically diverse habitats to range from US$0.23/ha to US$24.1/ha (in 2004 prices).

[3] A significant obstacle to developing a PES programme for water services is that the role of various land uses in generating water services is not very well understood. Land uses such as forest cover are widely believed to provide a variety of water services, but the evidence is often far from clear (Hamilton and King, 1983; Chomitz and Kumari, 1998; Calder, 1999; Bruijnzeel, 2004).

[4] In practice, PAs must submit requests to obtain the funds and there have been substantial delays in delivering them. Furthermore, IPAF funds are supposed to be spent only on PAs, but much of the degradation experienced in areas such as the Manupali River watershed lies outside PAs, and therefore is not covered by IPAF.

[5] For example, in southern Mexico the Scolel Té project is paying farmers to provide carbon sequestration services (Tipper, 2002). The Global Environment Facility (GEF) is supporting a pilot project aimed at using PES to generate biodiversity conservation and carbon sequestration benefits by encouraging the adoption of silvopastoral practices in degraded pastures in Colombia, Costa Rica and Nicaragua (Pagiola *et al.*, 2004). The World Bank's BioCarbon Fund is also developing approaches to land use-based carbon sequestration projects (Bosquet, 2004).

[6] Note that these figures exclude transaction costs associated with setting up and administering the PES programme. 1 t of carbon = 3.67 tCO_2e.

[7] This is also expressed in present value terms over 10 years, discounted at 12%. Note that payments need not be uniform in every year; indeed, front-loading payments may be necessary to induce adoption of forestry systems, in particular.

[8] For comparison, Costa Rica's PSA programme pays US$268/ha for conservation of existing forest and US$502/ha for reforestation (both over 10 years, discounted at 12%) (Pagiola, 2002).

[9] The cost could be further contained if the PES programme asked for bids to participate rather than offering a uniform payment. Given the relatively low difference in opportunity costs across high-input farms, however, the additional transaction costs may offset the gains in efficiency.

[10] Of course for water services, there may be little latitude in terms of participation: payments must go to whoever is in the upper watershed, whether private or public, small or large, rich or poor. Indeed, this is one reason why a PES programme may not necessarily benefit the poor: they may not be the ones controlling critical areas or resources of interest. Biodiversity services also may be quite site-specific.

References

Bosquet, B. (2004) *The BioCarbon Fund: An Overview*. The World Bank, Washington, DC.

Bruijnzeel, L.A. (2004) Hydrological functions of tropical forests: not seeing the soil for the trees? *Agriculture, Ecosystems and Environments* 104, 185–228.

Bulas, J.M. (2004) Implementing cost recovery for environmental services in Mexico. Paper presented at World Bank Water Week, Washington, DC, 24–26 February 2004.

Calder, I. (1999) *The Blue Revolution: Land Use and Integrated Water Resource Management*. Earthscan, London.

Chomitz, K. and Kumari, K. (1998) The Domestic Benefits of Tropical Forests: A Critical Review. *World Bank Research Observer* 13, 13–35.

Fondo Nacional de Financiamiento Forestal (FONAFIFO) (2000) El desarrollo del sistema de pago de servicios ambientales en Costa Rica. FONAFIFO, San José, Costa Rica.

Hamilton, L.S. and King, P.N. (1983) *Tropical Forest Watersheds: Hydrologic and Soils Response to Major Uses and Conversions*. Westview Press, Boulder, Colorado.

Hardner, J. and Rice, R. (2002) Rethinking green consumerism. *Scientific American*, May, 88–95.

Landell-Mills, N. and Porras, I. (2002) *Silver Bullet or Fools' Gold? A Global Review of Markets for Forest Environmental Services and Their Impact on the Poor*. International Institute for Environment and Development (IIED), London.

Lantican, M.A., Guerra, L.C. and Bhuiyan, S.I. (2003) Impacts of soil erosion in the upper Manupali watershed on irrigated lowlands in the Philippines. *Paddy Water and Environment* 1, 19–26.

Nelson, A. and Chomitz, K. (2002) The forest-hydrology-poverty nexus in Central America: an heuristic analysis. The World Bank, Washington, DC.

Pagiola, S. (2002) Paying for water services in Central America: Learning from Costa Rica. In: Pagiola, S., Bishop, J. and Landell-Mills, N. (eds) *Selling Forest Environmental Services: Market-based Mechanisms for Conservation and Development*. Earthscan, London.

Pagiola, S. and Platais, G. (forthcoming) Payments for Environmental Services: From Theory to Practice. The World Bank, Washington, DC.

Pagiola, S., Landell-Mills, N. and Bishop, J. (2002) Making market-based mechanisms work for forests and people. In: Pagiola, S., Bishop, J. and Landell-Mills, N. (eds) *Selling Forest Environmental Services: Market-based Mechanisms for Conservation and Development*. Earthscan, London.

Pagiola, S., Agostini, P., Gobbi, J., de Haan, C., Ibrahim, M., Murgueitio, E., Ramírez, E., Rosales, M. and Ruíz, J.P. (2004) Paying for biodiversity conservation services in agricultural landscapes. Environment Department Working Paper No.96. The World Bank, Washington, DC.

Pagiola, S., Arcenas, A. and Platais, G. (2005) Can payments for environmental services help reduce poverty? An exploration of the issues and the evidence to date from Latin America. *World Development* 33, 237–253.

Principe, P. (1989) *The Economic Value of Biodiversity Among Medicinal Plants*. OECD, Paris.

REECS (2003) *Developing Pro-poor Markets for Environmental Services in the Philippines*. International Institute for

Environment and Development (IIED), London, UK.

Reid, W.V., Laird, S.A., Gamez, R., Sittenfeld, A., Janzen, D.H., Gollin, M.A. and Juma, C. (1993) A new lease on life. In: Reid, W.V., Laird, S.A., Meyer, C.A., Gamez, R., Sittenfeld, A., Janzen, D.H., Gollin, M.A. and Juma, C. (eds) *Biodiversity Prospecting: Using Genetic Resources for Sustainable Development*. World Resources Institute, Washington, DC.

Rice, R. (2003) *Conservation Concessions: Concept Description*. Conservation International, Washington, DC.

Rola, A.C., Sumbalan, A.T. and Suminguit, N.J. (2004) Realities of the watershed management approach: The Manupali Watershed experience. Working Paper No.04-04. Institute of Strategic Planning and Policy Studies, Los Baños, Philippines.

Shively, G.E. (2006) Externalities and labour market linkages in a dynamic two-sector model of tropical agriculture. *Environment and Development Economics* (in press).

Simpson, R.D., Sedjo, R.A. and Reid, J.W. (1996) Valuing biodiversity for use in pharmaceutical research. *Journal of Political Economy* 104, 163–183.

Tipper, R. (2002) Helping indigenous farmers participate in the international market for carbon services: the case of Scolel Té. In: Pagiola, S., Bishop, J. and Landell-Mills, N. (eds) *Selling Forest Environmental Services: Market-based Mechanisms for Conservation and Development*. Earthscan, London, pp. 223–233.

Winrock International (2004) Financial incentives to communities for stewardship of environmental resources. Feasibility study LAG-A-00-99-00037-00. Winrock International, Washington, DC.

World Bank. (2004) State and trends of the carbon market 2004. The World Bank, Washington, DC.

Zelek, C.A. and Shively, G.E. (2003) Measuring the opportunity cost of carbon sequestration in tropical agriculture. *Land Economics* 79, 342–354.

12 Conclusions and Some Directions for Future Research

I. COXHEAD,[1] A. C. ROLA[2] AND G. SHIVELY[3]

[1]Department of Agricultural and Applied Economics, University of Wisconsin-Madison, 427 Lorch Street, Madison, WI 53706, USA, e-mail: coxhead@wisc.edu; [2]University of the Philippines at Los Baños, College, Laguna 4031, Philippines, e-mail: arola@laguna.net; [3]Department of Agricultural Economics, Purdue University, 403 West State Street, West Lafayette, IN 47907, USA, e-mail: shivelyg@purdue.edu

In this book we have assembled the results of a long-term, coordinated, mutidisciplinary set of research endeavours with two goals. The first is to provide a framework for studying economic and environmental impacts of economic and policy changes, especially as they impact upland agricultural producers and the communities in which they live. The second is to help build a prototype for natural resource and environmental research by academics, and planning by policymakers for whom natural resource management is an imperative. This goal we achieve by documenting, in the case of the Manupali River watershed, the forces leading to land use change and the potential impacts of policy changes and institutional changes.

Our inspiration has been the ground-breaking work of earlier developing-country studies focusing on the watershed, particularly those by Easter et al. (1986) and Doolette and Magrath (1990). Subsequent work, including ours, has deepened and sharpened the value of these important contributions. We have taken explicit account of two near-universal trends of the past 15–20 years: agricultural commercialization and the increasingly complete integration of upland farming communities into national and global markets for products, labour and capital; and the equally fundamental shift of administrative and policy powers from central control (in name if not in practice) to explicitly decentralized, and in some cases community-oriented, decision making. Our work has also benefited by unprecedented access to new analytical tools and above all, new data, especially remote sensing data, detailed time-series data on land use, forest, soil and water quantity and quality, and panel data on prices, crop choices and other land use decisions by upland farmers.

Summary of Findings

At the end of the Second World War, most sloping and high-altitude land in South-east Asia was still forested. Over the next half century, a combination of upland population growth, infrastructural investments, new commercial opportunities, weak and ineffective ownership and regulatory institutions all encouraged dramatic rates of deforestation and agricultural expansion. This, in turn, contributed to very rapid rates of decline in watershed function. The Manupali River watershed in the southern Philippines, the setting for the research highlighted in this book, exhibits a typical pattern of land use change and change in watershed function. In 1973, approximately 28% of the area of the watershed was agricultural land used in the cultivation of temporary crops and 50% was undisturbed forest. Over the next 20 years, as cultivated area expanded, the forested area shrank to slightly over one-quarter of the total area. At the same time, overall streamflow diminished while seasonal flow variability, sediment and other pollutant loadings, and susceptibility to extreme weather events such as El Niño conditions, all increased. In these respects, the experience of the Manupali River watershed is one that is shared with many tropical watersheds.

A number of economic phenomena have helped shape the observed patterns of land use. These include in-migration as a result of poor labour absorption in the general economy, and a strong policy bias in favour of annual crops such as grains and vegetables. On the commodity side, relative prices have changed in a discernible way over time, favouring annual crops such as maize and vegetables over less erosive perennial crops. Trends in input prices have also been influential, as the major crops differ considerably in their employment of factors. For these reasons, markets and national economic policies have been instrumental in determining land use changes.

Institutional failures, broadly defined, have also been influential. Poor definition of property rights and overlapping land claims have both undermined incentives to invest in forests, soils and watershed function. Mismatched mandates and powers between national government, local government and communities have inhibited the design, adoption and implementation of effective resource management strategies. Since 1992, decentralization has placed important policy instruments into the hands of local administrations whose capacity to deploy them is inadequate. Interjurisdictional externalities arising from the mismatch of administrative and geographic boundaries persist.

Since the mid-1990s, available evidence shows that deforestation rates in the Manupali River watershed have diminished. In part this trend reflects a decline in the availability of attractive land, since much potentially useful land has already been cleared. But the reduction can also be traced to two important changes: economic growth in the general economy that has induced net out-migration from the watershed, thereby reducing pressure on an existing resource base, and better cooperation between state and local communities to manage watershed resources. The future of resource management depends on a continuation of these trends, above and beyond any policy measures that may be required. In the policy arena, it remains true that the bias of economic policy still favours production of annual crops. New technologies and new forms of economic organization such as agribusiness enterprises pose additional challenges to traditional resource management policy models.

In Chapters 1 and 2 we characterized the evolution of upland economies as occurring over three phases. The first is the era prior to modern economic growth; the second is the period in which markets advance just as institutional efficacy retreats, starting a race for open access resources. Our study area, like many other upland economies, is now in transition to a third development phase. In this, alternate pathways seem to lead in the direction of distinct environmental outcomes. In the preferred outcome, institutions will evolve to catch up with the pressures brought about by economic development, creating the conditions for a balance between the demands of current economic activity and the need to conserve resources for the future. Along an alternate (and less preferred) path, institutional development lags so far behind economic development pressures that long-term growth is compromised by pervasive and possibly severe environmental damages. Will the third phase of economic development see moves towards sustainability, or environmental breakdown? In the Manupali River watershed signs of both trends are present. Our findings demonstrate that much will depend on trends in the external economic environment and in the local institutional and regulatory setting.

Lessons

How can the populations of upland watersheds attain the best long-term economic and environmental outcomes? To grapple with the challenges of watershed management in the Philippines and elsewhere it will be necessary to adopt a range of measures. Some possibilities highlighted in this volume include identifying and targeting efforts on environmental hotspots for the protection and rehabilitation of the watershed (Chapter 8), incorporation and adaptation of new farming methods and cropping practices (Chapter 9), serious consideration of agricultural pricing policies that do not undermine environmental goals (Chapters 6 and 10), and innovations to reward upland groups for environmental stewardship (Chapter 11).

Each of these individual measures implies a policy or programme that is primarily the responsibility of a single agency, project or group. Without doubt, however, viable solutions to the problems associated with current patterns of land and natural resource use will require effort on numerous fronts and involve all levels of government and stakeholders working together – or at least in coordinated fashion. From the experience of the Manupali River watershed, it is possible to derive some general insights for policy and institutional design, as follows.

Better accountability requires genuine decentralization

Genuine decentralization implies substantial local autonomy with respect to fiscal, administrative and political powers. From our case study, we see village- and community-level natural resource management institutions becoming increasingly viable (if from a low starting point), and greater willingness on the part of the provincial government to invest in environmental programmes (Sumbalan, 2001). However, at the very local levels (municipality or village), appreciation of the long-term benefits of good environmental management is still lacking, and furthermore, capacity building is needed to monitor and interpret environmental indicators. Care must therefore be taken to ensure that support for capacity building, connections to information and research networks, and other ancillary strengths accompany the transfer of formal powers. These are essential components of a genuine decentralization, that is, one that goes beyond merely transferring powers and mandates 'on paper' whilst retaining effective control over policy design and implementation at a higher level of government.

Addressing the externality problem requires watershed-based institutions and policies

In spite of the intrinsic desirability of decentralization, careful attention must also be paid to the definition of resource management institutions and to the assignation of functions at each level of governance. Given the many externalities involved, there is a compelling case for management of forest, land and water resources at the watershed or river basin level (Dixon and Easter, 1986; Francisco, 2002). This means defining the hydrologic unit and providing for an administrative structure for this level of management. Both the province of Bukidnon and the municipality of Lantapan have begun to create and to take part in forest and watershed management bodies with mandates that transcend political boundaries. However, strong cooperation among villages and across local government units is needed, especially in the regulatory and fiscal aspects.

Having defined the appropriate institutions, it is equally important that they be awarded the powers and the fiscal means to carry out their work. Adding new layers to the administrative and policy structure may be costly, but will yield nothing (or worse, have a negative result by undermining existing institutions without replacing them) unless they have the resources to fulfil their functions.

Market-based mechanisms can potentially support sustainable resource management

An important lesson from our study of the Manupali River watershed is that market expansion, on its own, is not the sole cause of unsustainable development. Property rights failures, externalities and incomplete markets are almost always present, and these serve to distort market-driven resource allocation and conservation decisions (Barbier, 1990; Pagiola *et al.*, 2002). First-best solutions require addressing failures at their source, and then designing policies for agri-cultural development. Uncontrolled market expansion without property rights or corrective and regulatory institutions results in patterns of development and land use in which resource depletion and environmental destruction is rapid, perva-sive and potentially disastrous.[1]

Market-based incentives can be also used by local governments to encour-age sustainable land use decisions. Subsidies or the kinds of payments high-lighted in Chapter 11 can be offered to farmers who agree to practise appropriate soil conservation measures, including tree-planting or buffer-zone management. In practice, there may be a tendency for participants to over-state cost in the face of potential subsidies (Goodstein, 1999), but incentive-compatible approaches can be developed to encourage truthful behaviour (Kwerel, 1977 as cited by Goodstein, 1999; Zelek and Shively, 2005).

New institutional arrangements require innovative approaches to regulation

New resource management institutions have the advantage that they transcend sectoral, geographic and administrative boundaries. But their disadvantage is that they rely mainly on 'command and control' measures. This creates (or more accurately, reinforces) tensions between regulations and market incen-tives, the former demanding behaviour that is incompatible with the latter. The next phase of institutional and regulatory reform will have to deal with this by designing and deploying market-based instruments.

Some tentative steps in this direction are now being taken. In the Manupali River watershed, one recently adopted municipal ordinance stipulates that if farmers practise soil conservation measures, they will be favoured as partici-pants in government programmes. Tree farming and agroforestry, when weakly promoted, may not take place on a sufficiently large scale to arrest ero-sion problems. One recent study underscores the importance of price policy intervention for tree-growing in promoting smallholder, tree-based farming sys-tems (Predo, 2002).

Directions for Future Research

As with any topic of this magnitude and complexity, many areas of research remain to be covered. One question that remains unresolved in this study is that of the role of intangible community assets such as social capital. In the context of an upland watershed, increased social capital, which implies stronger and 'thicker' networks of horizontal ties among agents, is seen to improve governance and can lead to increasingly cooperative community ~tion, thus helping to resolve problems of collective action and common property (Narayan and Pritchett, 1997). A literature is now beginning to emerge on the role of informal social institutions and network ties as con-tributing to a firmer basis for collective action and improved environmental outcomes (Fox and Gershman, 2000; Pretty and Ward, 2001). Broad-based participation in decision making, it is argued, is compatible with solutions to the collective action problem. It follows that future changes in management structures for upland resource management should include the lowlands and the broader economic environment. Adgar (2000) argues that social and ecological resilience may be mutually reinforcing, especially where communities are highly dependent on ecological and environmental resources.

On the other hand, divergent economic interests raise the cost of collective action and may inhibit good policy formation by broad-based forest management bodies such as the PAMB, or by a (notional, as yet) spatially broad watershed-based management authority. Within the Manupali River watershed, groups with strong social capital seem to provide better accountability to their members (Paunlagui *et al.*, 2003); however, the diversity of interests, ethnicities, wealth and geographic location all work against strong network formation and collective action within the municipality as a whole. It follows that future changes in management structures for upland resource management should strive to include means to resolve differences across these divisions, and in particular between upland and lowland agents.

Our work also brings out some methodological directions, both for researchers and for planners. Market integration, agricultural commercialization and capital deepening in upland economies are all significant and widespread trends. Relative to the quasi-autarkic state of subsistence farming, commercially oriented capital-intensive farm operations can both increase the rate at which natural resource services are exploited and also improve the efficiency of their use. Which tendency prevails is likely to depend in large part on property rights and regulatory regimes. In Lantapan, for example, commercial monoculture not only by specialized commercial family farms, but also and increasingly by large-scale plantations and intensive poultry and hog raising operations, has expanded greatly since the mid-1990s. Relative to the 'traditional' land use patterns of mainly subsistence-oriented farmers, the new uses of land and water resources produce very high private net returns; however, they also make major demands on soil and water resources and pose potentially large problems of water and air pollution. In this upland watershed, as in very many others across Asia, the next generation of challenges to sustainability may be driven much more by the decisions of larger commercial farms and agribusiness firms than by poor, semi-subsistence farm households. External markets, including those for labour and capital, can be expected to be increasingly influential. Finally, the continuing decentralization of authority and responsibility over natural resources management in most developing countries will bring about major changes from the old ways.

Local and national governments and other management agencies, as well as development projects, must adapt and innovate to meet these new challenges to old modalities. Future researchers must also begin to adapt their core questions, assumptions and methodologies to respond to this trend. The dynamic evolution of local economies and resource management decision mechanisms with 'new actors, and new roles for old actors' (Siamwalla, 2001), means that all models of watershed management, from the analytical to the practical and policy-oriented, will need to be highly flexible with regard to their characterization of institutional and management structures. Local awareness of the causes of land use change and local understanding of the economic and environmental implications of land use change are both necessary for improving policies that govern natural resource use. Equipped with such awareness and understanding, those empowered to alter institutions and policies that shape land use in tropical watersheds must then seek the participation of all relevant stakeholders, whether for purposes of policy design, implementation, enforcement or evaluation.

Note

[1] This is illustrated by such catastrophes as the 1992 Ormoc mudslides in the central Philippines, where denudation of a watershed created the underlying conditions that precipitated a landslide which destroyed an entire town.

References

Adgar, W.N. (2000) Social and ecological resilience: are they related? *Progress in Human Geography* 24, 347–364.

Barbier, E.B. (1990) The farm-level economics of soil conservation: the uplands of Java. *Land Economics* 66, 199–211.

Dixon, J.A. and Easter, K.W. (1986) Integrated watershed management: an approach to resource management. In: Easter, K.W., Dixon, J.A. and Hufschmidt, M.M. (eds) *Watershed Resources Management.* Westview Press, Boulder, Colorado.

Doolette, J.B. and Magrath, W.B. (1990) Strategic Issues in Watershed Development. In: Doolette, J.B. and Magrath, W.B. (eds) *Watershed Development in Asia.* World Bank Technical Paper No. 127. The World Bank, Washington, DC.

Easter, K.W., Dixon, J.A. and Hufschmidt, M.M. (eds) (1986) *Watershed Resources Management.* Westview Press, Boulder, Colorado.

Fox, J. and Gershman, J. (2000) The World Bank and social capital: lessons from ten rural development projects in the Philippines and Mexico. *Policy Sciences* 33, 399–419.

Francisco, H.A. (2002) Why watershed-based water management makes sense. PIDS Policy Notes 2002-09. Philippine Institute for Development Studies, Makati, Manila, Philippines.

Goodstein, E.S. (1999) *Economics and the Environment.* Prentice-Hall, Upper Saddle River, New Jersey.

Kwerel, E. (1977). To tell the truth: imperfect information and optimal pollution control. *Review of Economic Studies* 44, 595– 601.

Narayan, D. and Pritchett, L. (1997) Cents and sociability: household income and social capital in rural Tanzania. Policy Working Paper No. 1796. The World Bank, Washington, DC.

Pagiola, S., Landell-Mills, N. and Bishop, J. (2002) Making market based mechanisms work for forests and people. In: Pagiola, S., Bishop, J. and Landell-Mills, N. (eds) *Selling Forest Environmental Services: Market Based Mechanisms for Conservation and Development.* Earthscan, London.

Paunlagui, M.M., Nguyen, M. and Rola, A.C. (2003) Social capital, ecogovernance and natural resource management: a case study in Bukidnon, Philippines. ISPPS Working Paper No. 03–04, University of the Philippines at Los Baños, College, Laguna, Philippines.

Predo, C. (2002) Bioeconomic modeling of alternative land uses for grassland area and farmers' tree-growing decisions in Misamis Oriental, Philippines. Unpublished PhD dissertation, University of the Philippines at Los Baños, College, Laguna, Philippines.

Pretty, J. and Ward, H. (2001) Social capital and the environment. *World Development* 29, 209–227.

Siamwalla, A. (2001) *The Evolving Roles of State, Private and Local Actors in Rural Asia.* Oxford University Press for the Asian Development Bank, New York.

Sumbalan, A.T. (2001) The Bukidnon experience on natural resource management decentralization. Paper presented at the SANREM conference, ACCEED Makati, Philippines, May.

Zelek, C. and Shively, G. (2005) Asymmetric information and contract specification for environmental services. Unpublished manuscript. Purdue University, West Lafayette, Indiana.

Index

Page numbers for figures/tables appear in **bold** type
Page numbers for main entries which have subheadings refer to general/introductory aspects of that topic.